Hydrology: Advances in Theory and Practice

In Focus – Special Book Series

Hydrology: Advances in Theory and Practice

Editor
Nevil W. Quinn

IWA PUBLISHING

Published by **IWA Publishing**
Alliance House
12 Caxton Street
London SW1H 0QS, UK
Telephone: +44 (0)20 7654 5500
Fax: +44 (0)20 7654 5555
Email: publications@iwap.co.uk
Web: www.iwapublishing.com

First published 2020

British Library Cataloguing in Publication Data
A CIP catalogue record for this book is available from the British Library

ISBN: 9781789061420 (paperback)
ISBN: 9781789061437 (eBook)
ISBN: 9781789061444 (ePUB)

Contents

Editorial: Hydrology: advances in theory and practice

The British Hydrological Society (BHS) has a long tradition of scholarly meetings – the *First National Hydrology Symposium* was held at Hull University in September 1987. This Special Issue collates papers presented at the *Thirteenth National Hydrology Symposium* held on 12–13 September 2018 at the University of Westminster, London. In addition to these 13 national symposia, the Society convenes an International Conference every 6 years – the most recent of which was the *Fourth BHS International Conference* held at Cranfield University (30 August–1 September 2016) (see *Hydrology Research* **49**(2)).

The theme for the 2018 conference was *Hydrology: Advances in Theory and Practice*, a fitting title given that the membership of the Society is approximately 50% academic and 50% practitioners, and contributions to both the theory and practice of hydrology are evident in this Special Issue. Flood estimation methods usually feature prominently at BHS meetings, and on this occasion we have contributions to methodologies for extreme value estimation (Hammond 2019), and improvements to an important catchment descriptor (Griffin *et al.* 2019). The value of distributed hydrological modelling for decision making is highlighted (Ocio *et al.* 2019) and recent advances in efficient coupling of hillslope hydrology and hydraulics for natural flood management are reported (Hankin *et al.* 2019). West *et al.* (2019) highlight the importance of the North Atlantic Oscillation (NAO) in influencing the spatial distribution of monthly precipitation in the UK and Muchan & Dixon (2019) present important empirical work on raingauge undercatch. We were pleased to receive papers focussing on the hydrology of other regions too: Hudson & Thompson (2019) on modelling climate change impacts in Siberia's Lena River basin and Magliano *et al.* (2019) on the hydrological functioning of cattle ranching impoundments in Argentina.

Special mention must be made of the two papers in this issue which were also invited keynote presentations. Wilby (2019) presents a thought-provoking outlook on a global hydrology research agenda fit for the 2030s, and Beven (2019) is an equally challenging and insightful piece on how to make advances in hydrological modelling – based on almost 50 years of modelling experience. The British Hydrological Society is very grateful to IWA Publishing for waiving Open Access fees for these two excellent papers, thereby ensuring that they attract a global readership.

The symposium was also held in honour of Professor Geoff Petts (1953–2018), who as Past President of BHS (2015–2017) and convener of this symposium, had undertaken considerable planning for it before his illness and subsequent death in August 2018.

Guest Editor
Nevil W. Quinn
University of the West of England, UK

REFERENCES

Beven, K. 2019 How to make advances in hydrological modelling. *Hydrology Research* **50** (6), 1481–1494.

Griffin, A., Young, A. & Stewart, L. 2019 Revising the BFIHOST catchment descriptor to improve UK flood frequency estimates. *Hydrology Research* **50** (6), 1508–1519.

Hammond, A. 2019 Proposal of the 'extreme rank plot' for extreme value analysis: with an emphasis on flood frequency studies. *Hydrology Research* **50** (6), 1495–1507.

Hankin, B., Metcalfe, P., Beven, K. & Chappell, N. A. 2019 Integration of hillslope hydrology and 2D hydraulic modelling for natural flood management. *Hydrology Research* **50** (6), 1535–1548.

Hudson, C. E. & Thompson, J. R. 2019 Hydrological modelling of climate change impacts on river flows in Siberia's Lena River Basin and implications for the Atlantic Meridional Overturning Circulation. *Hydrology Research* **50** (6), 1577–1595.

Magliano, P. N., Mindham, D., Tych, W., Murray, F., Nosetto, M. D., Jobbágy, E. G., Niborski, M. J., Rufino, M. C. & Chappell, N. A. 2019 Hydrological functioning of cattle ranching impoundments in the Dry Chaco rangelands of Argentina. *Hydrology Research* **50** (6), 1596–1608.

Muchan, K. & Dixon, H. 2019 Insights into rainfall undercatch for differing raingauge rim heights. *Hydrology Research* **50** (6), 1564–1576.

Ocio, D., Beskeen, T. & Smart, K. 2019 Fully distributed hydrological modelling for catchment-wide hydrological data verification. *Hydrology Research* **50** (6), 1520–1534.

West, H., Quinn, N. & Horswell, M. 2019 Regional rainfall response to the North Atlantic Oscillation (NAO) across Great Britain. *Hydrology Research* **50** (6), 1549–1563.

Wilby, R. L. 2019 A global hydrology research agenda fit for the 2030s. *Hydrology Research* **50** (6), 1464–1480.

doi: 10.2166/nh.2019.001

A global hydrology research agenda fit for the 2030s

Robert L. Wilby

ABSTRACT

Global assessments show profound impacts of human activities on freshwater systems that, without action, are expected to reach crisis point in the 2030s. By then, the capacity of natural systems to meet rising demands for water, food, and energy could be hampered by emerging signals of anthropogenic climate change. The hydrological community has always been solution-orientated, but our generation faces perhaps the greatest array of water challenges in human history. Ambitious programmes of research are needed to fill critical data, knowledge, and skills gaps. Priorities include filling data sparse places, predicting peak water, understanding the physical drivers of mega droughts, evaluating hyper-resolution models, managing compound hazards, and adjusting water infrastructure designs to climate change. Despite the opportunities presented by big data, we must not lose sight of the deep uncertainties affecting both our raw input data and hydrological models, nor neglect the human dimensions of water system change. Community-scale projects and international research partnerships are needed to connect new hydrological knowledge with most vulnerable communities as well as to achieve more integrated and grounded solutions. With these elements in place, we will be better equipped to meet the global hydrological challenges of the 2030s and beyond.

Key words | agenda, data, hazard, hydrology, model, research

Robert L. Wilby
Department of Geography and Environment,
Loughborough University,
Leicestershire LE11 3TU,
UK
E-mail: *r.l.wilby@lboro.ac.uk*

INTRODUCTION

The future is much closer than it used to be. A decade ago, warnings about an approaching global food, energy, and water crisis were a cloud on the distant horizon. At that time, projections of economic and population growth by the 2030s suggested a 50% rise in the demand for food and energy, with a 30% increase in water requirements, while having to contend with the early signs of anthropogenic climate change (Beddington 2009). Other drivers of change to national security, social cohesion, economic prosperity, and environmental sustainability are also emerging. For instance, 'megatrends' in public debt, information and communication technology, urbanization, and the rise of the individual, all shape global patterns of natural resource consumption and create new vulnerabilities (KPMG 2014; Wilby 2017). Two thirds of the human population could experience progressive increases in drought conditions under global warming (Naumann *et al.* 2018). With greater interconnectedness of economies and supply chains, we are all potentially vulnerable, whether directly or indirectly, to extreme hydrological events.

Meanwhile, numerous global assessments are cataloguing the profound impacts of human activities on freshwater systems. These include measurable groundwater depletion across swathes of North Africa, North America, the Middle East, and South Asia (Rodell *et al.* 2018); widespread river regulation, watershed disturbance and contamination

doi: 10.2166/nh.2019.100

(Vörösmarty *et al.* 2010). Evolving and/or intensifying threats to freshwater biodiversity also come from changing climates; global internet commerce and spread of non-native species; infectious diseases; harmful algal blooms; expanding hydropower; emerging contaminants; engineered nanomaterials; microplastic pollution; light and noise; freshwater salinization; declining calcium; and cumulative stressors (Reid *et al.* 2019: 849).

As the 'perfect storm' of the 2030s forecasted by Beddington (2009) closes in, there appears to be greater urgency about tackling the climate–food–energy–water nexus through policy and planning at various levels (e.g. Bazilian *et al.* 2011; Conway *et al.* 2015; Yang *et al.* 2016). In 2015, the United Nations Sustainable Development Goals succeeded the Millennium Development Goals. The new global vision set targets for 2030 based on 17 themes that are highly interdependent and intrinsically water-related (Harmancioglu 2017). Goal 6 is to *ensure availability and sustainable management of water and sanitation for all.* Yet, more impetus has been added by the realization that global mean warming of 1.5 °C could be reached as soon as 2030 with concomitant risks to human and natural systems (IPCC 2018). Governments and institutions now find themselves under pressure to acknowledge a 'climate emergency' and to commit to much more ambitious mitigation pathways to achieve net zero emissions.

How should the hydrological community respond to such major imperatives? What new information and research are needed to support a concerted global effort to manage the water challenges of the 2030s and beyond? Traditionally, we have arranged our knowledge within thematic silos that have changed little over the last 25 years (Table 1). However, greater emphasis is now needed on preparing for hydro-system changes that could fall outside historic variability (Wagener *et al.* 2010). This Comment Paper makes a case for a more integrated, solution-orientated approach to the global water-related challenges identified above (following Carpenter *et al.* 2009; Pahl-Wostl *et al.* 2013; Green *et al.* 2017a). These thoughts were shared in a closing address to the British Hydrological Society National Meeting in September 2018, then updated to reflect later developments.

Research opportunities surrounding six solution sets are identified. These are based on the personal reflections of an applied hydroclimatologist with the good fortune of working in five continents over the last 30 years. A call is made for an agenda aimed at (1) filling data sparse places; (2) predicting peak water; (3) understanding the physical drivers of mega droughts; (4) evaluating hyper-resolution models; (5) managing compound hazards; and (6) adjusting engineering standards (under non-stationary conditions). These are followed by a section that reflects on ways of working together that strengthen multidisciplinary integration and research impact. Throughout, the emphasis is on supporting the most vulnerable communities. It is hoped that these suggestions will inform wider discourse about the future direction and priorities for hydrological science, including associated training needs. There is also scope for the closer alignment of our research and development programmes with complementary disciplines.

Table 1 | Major themes addressed by the fourth (1993) and thirteenth (2018) British Hydrological Society National Symposia

4th BHS National Symposium 1993	13th BHS National Symposium 2018
Water Resources	*Droughts, Low Flows and Resources*
Environmental Aspects	*Advances in Hydroecology*
	Advances in Sediments and Habitats
Information Systems	*Advances in Hydrometric Data*
Floods	*Flood Hydrology*
Modelling Developments	*Advances in Modelling*

FILLING DATA SPARSE PLACES

Quality assured hydrological data are needed to adaptively manage resources, calibrate remotely sensed observations, build, and test models. Nonetheless, there are vast tracts of Earth lacking ground measurements of fundamental water balance components (i.e. precipitation, evapotranspiration and changes in ice, lake/wetland, soil, or groundwater storage). Network densities are particularly sparse in the Arctic, sub-Saharan Africa (apart from South Africa), central Asia, the Pacific Islands, and South America. High-altitude regions and fragile states are especially under-represented. Even where there are data, records may be incomplete due to lack of resources for personnel and equipment or because of conflict; there may be homogeneity issues due to site, instrument, or observer changes; data may be corrupted at any point in the information flow or held in inaccessible formats (Wilby *et al.* 2017). For example, temporal variations in the rating curves used to relate channel cross-section, flow depth, and discharge are a source of uncertainty even for well-resourced agencies (see Slater *et al.* 2019).

A range of initiatives are needed to fill data gaps. Equipment can be installed in strategically important places – so-called sentinel locations – where hydrological changes are most likely to be detected (Fowler & Wilby 2010). For instance, there is evidence that high elevation sites are warming more rapidly than the global mean (Pepin *et al.* 2015), but this may only hold true up to 5000 m (Gao *et al.* 2018). A spring 2019 expedition established automatic weather stations on Mount Everest, including the highest station in the world (at 8,430 m on the Balcony of the Southeast Ridge) (Wilkinson 2019). Clearly, the logistical, physical, and technical challenges that must be faced in such environments are extraordinary. However, more of these data are urgently needed to deepen understanding of the behaviour of the Hindu Kush Himalaya (HKH) water towers that ultimately support the livelihoods, water, and energy needs of more than 2 billion people (Immerzeel *et al.* 2010).

It is generally accepted that some of the most data sparse regions are also the most vulnerable to hydroclimatic hazards – globally, data are least available where they are most needed. Temporal scaling can extrapolate sub-daily, even sub-hourly, precipitation intensities for engineering design, where daily data exist (e.g. Courty *et al.* 2019). Alternatively, geostatistical techniques might be used to blend fragments of *in situ* data with remotely sensed information (e.g. Wilby & Yu 2013), including proxies for hydrological variables (e.g. Najmaddin *et al.* 2017) to run conceptual and/or distributed hydrological models (e.g. Samaniego *et al.* 2011). Basic 'hot-spot' sensitivity analyses may then identify communities most at risk or inform the design of future hydrometeorological network expansions. Comprehensive assessments are also needed to compare temporal and spatial variations in the skill of various remotely sensed, re-analysis, model- and ground-based hydro-meteorological data (e.g. Sun *et al.* 2018; Pritchard *et al.* 2019). Even then, post-processing and bias correction may be required to convert gridded products to point data.

The development of high-resolution gravity-based methods could literally 'weigh' water changes at whole catchment scales, thereby perhaps even supplanting traditional river gauging techniques (e.g. Gouweleeuw *et al.* 2018). Community-wide efforts are needed to build hyper-resolution (~1 km) models for monitoring terrestrial water, energy, and biogeochemical cycles (Wood *et al.* 2011).

Considerable data assets are still held in paper archives that have yet to be digitized. Recent projects demonstrate how well-coordinated public engagement ('citizen scientists') can recover otherwise dormant data. For example, the digitization of 20 years of nineteenth century hourly weather data from the summit of Ben Nevis revealed near-zero atmospheric humidity episodes at high altitudes (Burt & Hawkins 2019). Similarly, the rediscovery of very long sleet and snowfall record for the greater London area by Murphy *et al.* (2019) showed that a widely accepted narrative of wetter-winters in England and Wales since 1766 is, most likely, an artefact of snowfall under-catch in the pre-1860s. The upward winter precipitation trend becomes insignificant when corrected for snowfall. Hence, even in data-rich regions, old data viewed through a modern lens can sometimes challenge deeply held convictions about hydrological change (or not).

PREDICTING PEAK WATER

There are widespread concerns about growing water scarcity and associated risks to global food security (Falkenmark 2013), including the vulnerability of the international trade in staple crops to hydroclimatic shocks (d'Amour *et al.* 2016). Peak non-renewable water occurs in systems where rates of groundwater pumping or contamination lead to a maximum of production followed by a decline (analogous to the concept of peak oil). This stage has already been passed in the aquifers of Ogallala and California's Central Valley, the North China Plains, and in Andhra Pradesh, Rajasthan, and Tamil Nadu, India (Gleick & Palaniappan 2010). One estimate suggests that present global groundwater withdrawals are equivalent to a footprint that is 3.5 times the actual area of the aquifers and that 1.7 billion people live in regions where groundwater resources are under threat (Gleeson *et al.* 2012: 197).

One third of humanity depends on water supplies from large glacierized catchments (Huss & Hock 2018). Hence, global shrinkage of glaciers and associated changes in meltwater and runoff will have profound implications for water supplies, irrigation, and the energy security of downstream communities. In parts of the Tien Shan, Central Asia, meltwater from glacial stores could culminate in the 2020s (Sorg *et al.* 2014). However, there is uncertainty about the exact timing of peak meltwater in most regions. This partly stems from a lack of long-term data as noted above. Peak water also depends on the size of the catchment and relative contributions of ice, snow, thawing permafrost, and rainfall to total flow. Climate model and emissions scenario uncertainty matter too, especially by the end of the century when projections diverge according to the mitigation pathway followed (e.g. Lutz *et al.* 2013; Shannon *et al.* 2019).

Overall, peak meltwater is expected to occur later in larger catchments (Huss & Hock 2018) and on the eastern (rather than western) slopes of the Tien Shan-Pamir-Karakoram mountain complex (Luo *et al.* 2018). Dynamical ablation models tend to show peak meltwater in later decades than mass balance models, because they can sustain snow and ice flux from higher elevations to the melt zone for longer, rather than simply ablating as a static block (Kure *et al.* 2013; Ohara *et al.* 2014). However, both types of ablation model are confounded by large deficiencies in climate model representations of physical processes over mountainous regions, including temperature lapse rates, land surface feedbacks, and precipitation (Dobler *et al.* 2012). More research is needed to reconcile differences between glacier melt models within the significant constraints of input data and parameter uncertainty.

Although observed ice loss in the western Himalayas has accelerated over the last 40 years, it is too simplistic to attribute such trends entirely to rising temperatures; atmospheric deposition of black carbon (sooty particulates) and regional precipitation changes are also important drivers (Lutz *et al.* 2014; Maurer *et al.* 2019). Reliable projections of meltwater contributions from individual basins are hindered by the lack of information on seasonal snowpack, ice and hydrometeorological changes at high elevations, a partial understanding of the significance of black carbon and debris cover to ice melt, and by low confidence in climate model projections of precipitation changes, especially over the HKH. There are opportunities to develop robust data gathering techniques aimed at calibrating models, such as extracting snowline observations from remote imagery as a mass balance indicator (e.g. Barandun *et al.* 2015). More information on precipitation changes above 5,000 m is urgently needed, as well as parsimonious ablation models that include important processes such as sublimation (e.g. Wimmer *et al.* 2009).

The timing of peak water may be uncertain, but earlier, more rapid melt of snowpack followed by diminished summer flows is widely observed and consistently produced by model simulations under rising air temperatures. Seasonal changes in the timing and volume of meltwater have implications for the safety and operation of downstream hydropower facilities. Hence, procedures are being developed to seasonally forecast snowpack runoff to support the management of reservoir storage (Archer & Fowler 2008; Dixon & Wilby 2016; Apel *et al.* 2018). Forecast skill rests on links between leading modes of climate variability, such as ENSO, and regional hydrological hazards (Emerton *et al.* 2017). Lagged relationships between teleconnection patterns, winter precipitation, and spring snowmelt can then be exploited for inflow forecasting (e.g. Dixon & Wilby 2019). Generally, modest levels of skill mean that such techniques are best regarded as outlooks of above, near, or below average conditions, rather than precise forecasts. Further work is needed to account for non-stationarity in predictability, due to Arctic amplification causing poleward movement of the jet stream; expansion of the Hadley cell; interactions between climate modes; or variations in peak amplitude, duration, timing, and spatial patterns of sea surface temperature (SST) anomalies.

UNDERSTANDING THE PHYSICAL DRIVERS OF MEGA DROUGHTS

During a recent water management class, an undergraduate student asked *why* the Atlantic Meridional Oscillation (AMO) varies with a typical periodicity of 60–80 years. [If anyone has the answer, please let me know.] The question is perceptive because the AMO is linked to multi-decadal variations in the climate of northwest Europe (Sutton &

Dong 2012). During warm (cold) phases, summer rainfall is higher (lower) than average. Hence, anticipation of the AMO suggests foresight of rainfall anomalies – an expected return to cooler Atlantic SSTs in the 2020s and 2030s could imply a greater likelihood of severe European summer droughts (as in the 1970s and 1980s). Such information is highly pertinent to the UK water industry, given that the latest guidance requires companies to evaluate contingencies for a challenging but 'plausible worst case' drought that could exceed the coping capacity of existing and planned water supply systems (Environment Agency and Natural Resource Wales 2018).

Specification of a plausible extreme (mega) drought for resilience testing – here with notional 200 year return period – is a technically demanding task. Although data on mega drought occurrence have improved thanks to multi-centennial reconstructions (e.g. Cook *et al.* 2015; Wilby *et al.* 2015), higher-resolution information is needed for water resource systems modelling. This gap is being filled by stochastic weather generators and extreme value analysis, but there are concerns about the physical realism of the underlying models. The physical drivers of drought form a continuum from the seasonal dry-spell (storm track variation), through the multi-season episode (blocking), to the multi-year severe event (ENSO) and the drought-rich decade (AMO) (Pulwarty & Sivakumar 2014). So, is it reasonable to assume that a 2-year return period event can be drawn from the same population as a 200-year mega-drought?

Studies of persistent droughts in Australia suggest different processes behind the Federation (~1895–1902), World War II (1937–1945) and Millennium (2001–2009) events (Verdon-Kidd & Kiem 2009). Moreover, the various causes of drought could interact in ways that yield even more severe events (van Dijk *et al.* 2013). Multiple (not just single) modes of climate variability could also be driving severe droughts across Europe. For instance, Ionita *et al.* (2012) link inter-annual to decadal variability in European summer drought to three modes of SST anomalies; a long-term trend associated with warming over all oceans; interactions between the inter-annual ENSO and the Pacific Decadal Oscillation; and the AMO. Similarly, Folland *et al.* (2015) suggest that multi-annual droughts in the English lowlands could be linked to combinations of ENSO, North Atlantic SSTs, the Quasi-Biennial Oscillation, solar and volcanic forcings, and the AMO. More dynamical ocean–atmosphere modelling of multiple, interacting drivers is needed to uncover the causes of persistent rainfall deficits in the past and to explore how these might change in the future. Once a multi-year drought ends, even the rainfall-runoff relationship can change (Saft *et al.* 2015), implying that historical yields will not be reliable indicators of future water availability under climate change (Saft *et al.* 2016).

Hydrologists and climatologists also have shared interests in how soil-moisture feedbacks might amplify the severity of heatwaves and droughts under climate change. It is recognized that precipitation/soil moisture biases in climate models can lead to unrealistic partitioning of surface energy fluxes between land and atmosphere. For example, a dry bias in summer rainfall would lead to overly strong warming of extreme temperatures. Hence, by constraining climate model ensembles with observed correlations between summer rainfall and temperature, it is possible to narrow the range of uncertainty in future projections by excluding improbable models (Vogel *et al.* 2018). Elimination of such outlier (hot) models also reduces the projected multi-model median. Various global constraints on future changes in the hydrologic cycle have been considered, including tropospheric humidity, equilibrium, and transient mean precipitation (Allen & Ingram 2002), as well as the assumed mass balance between evaporation and precipitation (Liepert & Previdi 2012; Liepert & Lo 2013). Further work is needed to identify physically consistent, observation-based constraints to uncertainty in regional hydrological projections that are analogous to tests of climate model realism (e.g. McSweeney *et al.* 2015). A deeper understanding of land surface feedbacks could also be deployed in the design of green and blue infrastructure to counter urban heatwaves (e.g. Žuvela-Aloise *et al.* 2016; Gunawardena *et al.* 2017).

EVALUATING HYPER-RESOLUTION MODELS

The term 'hyper-resolution' was used earlier in relation to global water cycle monitoring at km-scales. The resolution of hydrological models is improving across all environments, assisted by progress in remote sensing, including low cost satellite systems and drone-based technologies (National Academies of Sciences Engineering and Medicine 2018). Here, for illustrative purposes, the focus is on urban flood simulation models operating at ~1–2 m horizontal and ±0.25 m vertical resolution at city-scales (e.g. Yin *et al.* 2015; Green *et al.* 2017b). Such models use rainfall forecasts to generate street-level maps of surface water flooding (Henonin *et al.* 2013) or to identify hotspots of vulnerability in terms of poor accessibility by emergency responders during flood episodes (e.g. Coles *et al.* 2017). Nowcasting applications assimilate rainfall forecasts every few minutes to update simulations of flood depth and extent. This raises demanding questions about how to appropriately test distributed models, given the large uncertainty in initial conditions, inputs, parameters, and outputs? Such challenges are common to all models of natural systems (Oreskes *et al.* 1994).

Beven (2019a, 2019b) proposes that model evaluation should be approached as a form of hypothesis testing. However, rather than test in a formal statistical sense, a case is made for assessing 'fitness-for-purpose'. For an operational flood forecasting system, this means appraising model behaviour under assumed boundary conditions. The fitness of a property-level forecast can be evaluated against a wide range of data seized *post hoc* (e.g. social media reports of flooding, recorded flood heights on buildings, street furniture or vehicles, insurance claims, remotely sensed data, and aerial surveys). For instance, Yu *et al.* (2019) compare distributions of simulated pluvial flooding across London with emergency responder data on the day of the EU referendum floods. Likewise, Muthusamy *et al.* (2019) combined high-resolution inundation data collected by a drone with property-level damage data for Cockermouth, Cumbria to evaluate fluvial plus pluvial flood simulations of storm Desmond in 2015. This required special permission for emergency response by the Civil Aviation Authority to fly over congested areas, beyond the visual line of sight and under extreme weather conditions. Understandably, such data sets are relatively rare, so there is scope to build libraries of data for model evaluation that could be shared among research teams.

Regardless of the domain of hyper-resolution hydrologic modelling (whether global- or street-level), there will always be a need for scale-dependent parameterization of sub-grid processes (Beven *et al.* 2015). Even at the scale of a high-resolution flood model, there is sub-grid heterogeneity in topography, soil properties, land cover, and vegetation that determine local infiltration rates, storage volumes, and flow pathways. In addition, precipitation amounts are typically interpolated between gauges to the grid-scale, whereas buildings and roads are taken as a snapshot in time. Hence, knowledge of these non-continuous land-surface and meteorological properties is subject to considerable epistemic uncertainty. How best to represent these variations at the catchment scale remains a fundamental research area.

In the meantime, highly localized model output can be made available to stakeholders who have the resident knowledge needed to identify errors (Beven *et al.* 2015). For example, emergency responders will no doubt be aware of the parts of the road network that regularly flood. Confidence star ratings can be applied to model representations of flood depth/area along with critical segments. Combined model uncertainty may also be represented as probabilistic visualizations of maximum inundation area for specified annual exceedance probabilities (Leedal *et al.* 2010). Given the danger of inconsistent analysis and messaging, there have been repeated calls for Codes of Practice to formalize the treatment of uncertainty (Pappenberger & Beven 2006), communication of flood (Demeritt & Nobert 2014), and drought risk (Climate Outreach 2016). Research is still required on how best to convey highly uncertain model outputs to different stakeholders, in consistent ways, recognizing that preferred formats will be context- and decision-maker-dependent.

MANAGING COMPOUND HYDRO-HAZARDS

Some of the most deadly and costly hydrological catastrophes are due to the coincidence of hazards in space or time. Traditional risk assessments viewed hazards one at a time; now, it is recognized that the likelihood of very high-impact events arising from combinations of climate drivers and/or hazards may have been underestimated (Hillier *et al.* 2015). For example, summer atmospheric blocking systems near Europe can produce extreme heat, drought, wildfire, and poor air quality as witnessed during the deadly 2010 Russian heatwave (Shaposhnikov *et al.* 2014). The joint occurrence of storm surge and fluvial and coastal flooding is perhaps the most familiar compound hydro-hazard as suffered during Hurricane Sandy, USA 2012, Typhoon Haiyan, Philippines 2013, and Cyclone Idai, Mozambique 2019. Other threats from hitherto rare hazard combinations, such as extreme heat following a hurricane, are only beginning to be investigated (Matthews *et al.* 2019). Whether through rising ambient temperatures, higher mean sea levels, more intense rainfall, or severe storms, climate change is generally expected to increase the likelihood of compound hazards (Wahl *et al.* 2015; Zscheischler *et al.* 2018).

Compound hazard definitions also include events with near temporal coincidence but manifested at separate locations. In addition, there are cascading disasters that arise from a single hazard which then triggers a chain of events resulting in large-scale impacts on lives and livelihoods (Cutter 2018). An example of the former is widespread, multi-basin fluvial flooding as witnessed in England and Wales during autumn 2000. At the height of the floods, nearly 20% of the drained area of these countries was recording near simultaneous annual maximum flows (De Luca *et al.* 2017). Moreover, peak flows tend to follow very severe gales, resulting in spatially distributed yet near coincident wind and flood damage. An example of a cascading hazard would be the extreme rainfall intensities in summer 2016 which led to flash-flooding and severe debris flow impacts on the small rural town of Braunsbach, Germany (Bronstert *et al.* 2018). This study highlighted the challenges of quantifying the multiple and cascading drivers behind the extreme event. Another illustration would be the link between heavy rainfall, urban surface water flooding, and cholera outbreaks in Senegal (de Magny *et al.* 2012).

Compound hazard analysis suggests that worst case years used for insurance purposes in catastrophe models are likely to be more costly than previously thought. There may also be 'hot-spots' where different types of hydrological extreme tend to recur at the same place (Collet *et al.* 2018). The UK National Flood Resilience Review and the UK Climate Change Risk Assessment (both in 2016) called for improved resilience to failures of interdependent critical networks (e.g. electricity, information technology, and transport) in high-risk locations. Taken together, there is a need to deepen understanding of the interrelationships between hazards and natural/human-induced environmental change. For instance, Gill & Malamud (2017) systematically identify 18 anthropogenic process types with 64 interactions that could potentially trigger/influence hazards. Hydrological hazards include vegetation removal or road construction increasing the susceptibility of slopes to landslide after heavy rain and or groundwater abstraction leading to more depressed river flows during a drought. Their matrices show the value of integrating anthropogenic processes within multi-hazard frameworks for more holistic location-specific screening of factors and thereby, disaster risk reduction.

As recognized by a 2019 call by UK Research and Innovation, there are many open questions around (1) the underlying drivers of compound hazards, (2) mechanisms of the cascade (or 'risk contagion') between drivers and receptors, as well as (3) options for improving resilience and managing multi-hazards. More research needs to be directed at low- and middle-income countries where there are relatively high systemic risks, and socio-economic vulnerabilities to compound hazards. To support national assessments, new spatially consistent analytical frameworks will be required to account for variations in hydroclimatic modelling uncertainty (e.g. Visser-Quinn *et al.* 2019). Ultimately, such developments should lead to improved hazard forecasting, civil contingency planning, and avoided damages.

ADJUSTING ENGINEERING STANDARDS

Climate change is expected to intensify extreme rainfall and raise global mean sea level so, without adaptation, damage from fluvial, pluvial, and coastal flood risk are expected to rise (Hirabayashi *et al.* 2013; Hallegatte *et al.* 2013). There are many enabling measures for managing these risks, such as forecasting, contingency planning for disasters, insurance, and land use zoning to reduce exposure. Site-specific interventions include new flood defence assets, upgrading resistance and resilience of existing infrastructure, modifying operating rules of flood control reservoirs, retreating from hazardous areas, periodic review, and adaptive management (Hallegatte 2009; Wilby & Keenan 2012).

Perhaps the most technically demanding option surrounds the adjustment of engineering standards to reflect evolving and projected hydrological conditions. This is especially contentious because of the methods of economic discounting applied to costs and benefits, as well as the low confidence in regional climate projections over the design life times of new infrastructure (Kundzewicz & Stakhiv 2010). New ways of working with non-stationary information must also be deployed (Serinaldi & Kilsby 2015).

Nonetheless, a few agencies are already providing look-up tables and guidance for engineers that reflect climate change (e.g. New York City Panel on Climate Change 2013; United States Army Corps of Engineers 2013; Netherlands Ministry of Infrastructure and Environment 2014; Asian Development Bank 2018; International Hydropower Association 2019). Some refer to 'adjustments', others to 'allowances,' or 'flood-risk reduction standards'. Some guides are intended to shape asset design, others for sensitivity (or 'stress') testing the performance of options. Latest advice on adapting to climate change in England provides tables of upper, central, and lower allowances for extreme rainfall intensity, peak river flow, and relative mean sea level, by region and period (2020s, 2050s, and 2080s) (Environment Agency 2016). Others advocate climate change allowances based on catchment type (Broderick *et al.* 2019) or precipitation mechanism/duration (Fowler & Wilby 2010). Ideally, global standards would emerge such that new infrastructure is built using consistent methodologies while respecting the deep uncertainties and regional variations in hydroclimatic and geotechnical risks. Even then, there may be a reluctance to adopt guidance depending on local appetite for risk and/or availability of resources.

Nonetheless, translation of climate model information into engineering standards is a non-trivial matter and can introduce considerable methodological uncertainty. First, judgements must be made about which emission(s) scenario, climate model ensemble(s) and part(s) of the ensemble range should be used. This depends on how precautionary the design must be, which eventually affects the cost of a project. Second, design variables must be extracted from climate model archives then post-processed to give the index severity, duration and return period as mandated by national design standards for specified structures. This involves decisions about whether to apply bias corrections to the climate model information to better match local measurements, as well as choices about the extreme value distribution. More elaborate analysis is required to derive sub-daily statistics from daily climate model output (e.g. Herath *et al.* 2016). Third, the resulting 'change factors' have to be expressed with reference to an agreed baseline, then aggregated spatially and rounded mathematically. This is necessary to avoid the impression of undue precision. Finally, guidance material and worked examples must be developed to help practitioners apply the tables correctly and consistently, while allowing audit by competent authorities. Ideally, the whole framework of activities would be transparent and subject to periodic review, in line with evolving knowledge.

The above description is an abridged version of a typical protocol for adjusting individual project designs. Ideally, these elements are arranged within an adaptive management framework that comprises a mix of hard, soft, and environmental measures. Regional- to national-scale programmes – such as the Thames Estuary 2100 (Ranger *et al.* 2013) and the Netherlands Delta Commission (Katsman *et al.* 2011) – involve lengthy, deliberative processes that engage with multiple stakeholder interests. Other examples of adaptive management pathways, include sequencing of public water supply augmentation options in Adelaide (Beh *et al.* 2015) and long-term water management in the Rhine Delta (Haasnoot *et al.* 2013). Although this strategy is appealing, highly contested socio-ecological trade-offs can be encountered, such as how to reallocate scarce water from irrigators to environmental flows. This is because the required amount of water to achieve an agreed benefit is uncertain even for individual species under present climate conditions, let alone for an entire freshwater ecosystem under climate change (Gell *et al.* 2019). A yet greater challenge is how to implement allowances for climate change where there are limited baseline data, low technical capacities, and/or weak governance.

WORKING TOGETHER

What we study is important but so too, are *how, where and with who* we research. This penultimate section offers a few suggestions about ways of working together that strengthen the multidisciplinary integration and impact of hydrological research.

To begin with, it is helpful to examine motivations. As elaborated before, there are many grand challenges, so every hydrologist has a stake in what happens next. However, there is a need for circumspection and realism about what individuals can contribute to solving significant global issues. Adams et al. (2015: 52) share helpful advice about ethical ways of working that include principles of integrity, transparency, humility, and collaboration. Their definition of humility is modified slightly to *presenting [ourselves] as no more or less than [we are], not promising more than can be delivered, nor obscuring an underlying reality of uncertainty*. Jennings *et al.* (2009) similarly provide views on the moral values and obligations set before everyone working in water resource management. Their ethical code stems from a sense of shared purpose, of working with nature, and of balance between traditional and new technologies. Their motivations are driven by care for the security, safety, and shared interests of both people *and* the environment.

Calls for closer integration of hydrology with other disciplines (especially social sciences and ecology) have been made before (e.g. Hannah *et al.* 2007; Krause *et al.* 2011; Sivapalan *et al.* 2012; Baldassarre *et al.* 2013; Montanari *et al.* 2013; Bierkens 2015; Di Baldassarre *et al.* 2015). We work in a dynamic landscape of research questions that are tackled by a variety of disciplines, and in which scientific findings are forging new disciplinary configurations (Vugteveen *et al.* 2014). For example, some claim that the ecological degradation of urban streams can only be reversed by integrating ecological research with the social, behavioural, and economic investigation (Walsh *et al.* 2005). Likewise, hydroclimatology is an emerging co-discipline that has brought valuable insights about the atmospheric drivers and land-surface conditions behind extreme hydrological events and their long-term behaviour at catchment to global scales (McGregor 2019). An alliance of hydroclimatology with hydrogeomorphology further offers a framework for interpreting links between modes of climate variability, river channel, and fluvial habitat changes (Slater *et al.* 2019). This land–water–eco-management nexus has stimulated much research on systems modelling and decision support. However, there are still opportunities for deeper integration of human–water dimensions in thematic areas such as sustainable development, hazards management, public–private ownership, and governance (Bakker 2010; Xu *et al.* 2018; Gell *et al.* 2019). In short, solution-orientated hydrology must treat human activities as *endogenous* to water system dynamics (Gober & Wheater 2015).

Multidisciplinary water research and knowledge exchange can take various forms, but genuinely co-productive research design is always a good starting point. In this way, the nascent research team negotiates a set of research questions that are meaningful to the practitioner and scientist alike – ideally, the communities served have a voice too. For instance, community champions can be invaluable sources of local knowledge for investigating flood impacts and coping strategies within low-income neighbourhoods (Gough *et al.* 2019). Early participation of such parties is essential for effective dissemination and uptake of research results as well. Some refer to this style of working as a bottom-up, resource-based vulnerability perspective (Pielke *et al.* 2012). The emphasis is very much on understanding contextual vulnerability and the quest for effective risk communication and low-cost prevention strategies. Conversely, top-down assessments typically involve macro-scale modelling of physical systems, considering multiple scenarios and some (but never all) dimensions of uncertainty. These are not mutually exclusive paradigms – the latter often frames the risk and international dimensions within which the former local solutions must ultimately reside (Conway *et al.* 2019).

Peer-to-peer collaborations can be highly productive in terms of two-way knowledge exchange and personal development. However, international partnerships face a range of barriers including narrow funding rules set by national agencies, inconsistent review processes, restricted access to facilities and/or data, intellectual property rights, and issues around cyber-security (Suresh 2012). Ideally, research council policies and incentives would promote mobility and interconnectedness of (early career) researchers across borders. The benefits could be wider participation in a global enterprise of hydrological problem-solving, regardless of uneven patterns of national science funding or research capacity.

Finally, there really is no substitute for fieldwork whether the purpose is to better formulate a research question, observe hydrological phenomena, or to collect new data. Unfortunately, in an age of open data, we can all become divorced from the processes involved in the gathering and scrutiny of primary information. As data sets are amalgamated, assimilated and/or post-processed, any errors and biases may become harder to detect (Wilby *et al.* 2017). Others are concerned about the opportunity costs of collecting data – when this time and resource could be devoted to analysing 'free' data and writing papers. The 'global perspectives' or 'global relevance' emphasized by some top journals may further disincentivise the pursuit of local field studies in hydrology. Hence, there may be a tension between the needs of the individual researcher and those of the wider community for new observations (Allen & Berghuijs 2018). Likewise, capacity development and knowledge exchange activities can be highly resourced intensive for academic hydrologists, but such services are an essential part of building more informed communities of practice (Watts 2015). Without such attention, there is a danger that researchers in countries with less capacity will be 'left behind' as new global datasets, remote sensing products, and hyper-resolution models become increasingly mainstream (Conway, personal communication). In summary, developing research collaborations, gathering new data, and investing time in people are all ways in which we can contribute to the hydrological mission.

CONCLUSIONS

This Comment Paper calls for a hydrological research agenda that is focused on solving the mounting challenges of global water, food, and energy security. The priority areas inevitably reflect the views of the author, so some critical knowledge gaps are likely under-represented. For instance, more weight could be given to tackling the global pandemic of arsenic poisoning by naturally contaminated groundwater (Chakraborti *et al.* 2003); or to the manifold threats faced by freshwater biodiversity (Reid *et al.* 2019), especially in headwaters (Riley *et al.* 2018); or to the uncertain regional hydrological ramifications of deforestation/afforestation combined with CO_2 fertilization (Prudhomme *et al.* 2014); or to stresses on transboundary water security (Siegfried *et al.* 2012), including the disruption of international flows of embodied water in commodities (Hunt *et al.* 2014). Some regard excessive nutrient enrichment and poor water quality as critical threats to ecosystem functioning (Woodward *et al.* 2012). Others are more intent on solving scientific problems in hydrology (e.g. Blöschl *et al.* 2019).

Additionally, there are the non-trivial technical challenges associated with the early detection and attribution of hydrological change. As well as reliable information about co-drivers of change, we also need conceptual and modelling frameworks that allow robust testing of multiple working hypotheses, such as the relative role play by climate, land use, and water management (e.g. Merz *et al.* 2012; Harrigan *et al.* 2014). Even then, statistically significant hydrological trends may not be detectable for decades (or even centuries to come) in environments with marked variability yet relatively weak signals of change (Ziegler *et al.* 2005; Wilby 2006). How might this be achieved in places with little or no data? How much change is *practically* (as opposed to statistically) significant? How does the likelihood of detection and attribution vary with catchment characteristics and/or choice of hydrological index? How should we respond when a credible signal emerges? Such cross-cutting questions warrant thematic programmes of inquiry in their own right.

The research agenda offered here is a multidisciplinary endeavour involving the expertise of hydrologists, climatologists, social scientists, ecologists, and many others, alongside stakeholders (Wheater & Gober 2015). There are also synergies with the science challenges identified by related communities (e.g. BGS 2019). Six priority themes were proposed here: (1) filling data sparse places; (2) predicting peak water; (3) understanding the physical drivers of mega droughts; (4) evaluating hyper-resolution models; (5) managing compound hazards; and (6) adjusting engineering standards (under non-stationary conditions). These will require the expansion of skills in hydro-informatics, data recovery, and visualization, while reinvigorating field observation, focusing on extreme hydrological events or hostile/remote locations.

We have unprecedented amounts of information and computing power at our disposal (Chen & Han 2016; McCabe *et al.* 2017). The Internet of Things offers scope for improved monitoring and modelling with adaptive management of water resources and hydrological hazards (e.g. Qiuming *et al.* 2012; Fang *et al.* 2015; Zhang *et al.* 2018). Such tools present exciting opportunities for fusing data streams from different sources and mining content for predictability. Yet, even in an era of 'big data', there are sensitive locations that remain data sparse. Many of these places are experiencing some of the most rapid hydrological change and/or are home to some of the most vulnerable people – priority environments are the global water towers (Immerzeel *et al.* 2010) and low-income communities of the global south (Douglas *et al.* 2008). Neither should we forget the deep uncertainties affecting our raw input data and hydrological models (Wilby *et al.* 2017; Ekström *et al.* 2018; Beven 2019a), nor the human dimensions to water system changes. With these elements in place, we will be better equipped to meet the unprecedented hydrological challenges of coming decades.

ACKNOWLEDGEMENTS

This commentary is dedicated to the memory of Geoff Petts. In our last correspondence, he invited me to share thoughts about directions for research in hydrology that address key global challenges. Hopefully, Geoff's care and foresight have been honoured. I am also indebted to Keith Beven, Declan Conway, Hayley Fowler, Harriet Orr, Murray Peel, and Howard Wheater for all their generous feedback on earlier drafts of this paper.

REFERENCES

Adams, P., Hewitson, B., Vaughan, C., Wilby, R. L., Zebiak, S., Eitland, E. & WMO Secretariat 2015 Call for an ethical framework for climate services. *WMO Bulletin* **64**, 51–54.

Allen, S. T. & Berghuijs, W. R. 2018 A need for incentivizing field hydrology, especially in an era of open data: discussion of 'The role of experimental work in hydrological sciences–insights from a community survey'. *Hydrological Sciences Journal* **63**, 262–1265.

Allen, M. R. & Ingram, W. J. 2002 Constraints on future changes in climate and the hydrologic cycle. *Nature* **419**, 228–232.

Apel, H., Abdykerimova, Z., Agalhanova, M., Baimaganbetov, A., Gavrilenko, N., Gerlitz, L., Kalashnikova, O., Unger-Shayesteh, K., Vorogushyn, S. & Gafurov, A. 2018 Statistical forecast of seasonal discharge in Central Asia using observational records: development of a generic linear modelling tool for operational water resource management. *Hydrology and Earth System Sciences* **22**, 2225–2254.

Archer, D. R. & Fowler, H. J. 2008 Seasonal forecasting of runoff on the River Jhelum, Pakistan, using meteorological data. *Journal of Hydrology* **361**, 10–23.

Asian Development Bank 2018 *Adjusting Hydrological Inputs to Road Design for Climate Change Risk Based on Extreme Value Analysis.* Report ADB PPTA 8957-VIE. Asian Development Bank, Manila, Philippines.

Bakker, K. 2010 *Privatizing Water: Governance Failure and the World's Urban Water Crisis.* Cornell University Press, Ithaca, NY.

Baldassarre, G. D., Viglione, A., Carr, G., Kuil, L., Salinas, J. L. & Blöschl, G. 2013 Socio-hydrology: conceptualising human-flood interactions. *Hydrology and Earth System Sciences* **17**, 3295–3303.

Barandun, M., Huss, M., Sold, L., Farinotti, D., Azisov, E., Salzmann, N., Usubaliev, R., Merkushkin, A. & Hoelzle, M. 2015 Re-analysis of seasonal mass balance at Abramov glacier 1968–2014. *Journal of Glaciology* **61**, 1103–1117.

Bazilian, M., Rogner, H., Howells, M., Hermann, S., Arent, D., Gielen, D., Steduto, P., Mueller, A., Komor, P., Tol, R. S. & Yumkella, K. K. 2011 Considering the energy, water and food nexus: towards an integrated modelling approach. *Energy Policy* **39**, 7896–7906.

Beddington, J. 2009 Food, energy, water and the climate: a perfect storm of global events. In: *Sustainable Development UK Annual Conference. QEII Conference Centre*, London. Government Office for Science.

Beh, E. H., Maier, H. R. & Dandy, G. C. 2015 Adaptive, multiobjective optimal sequencing approach for urban water supply augmentation under deep uncertainty. *Water Resources Research* **51**, 1529–1551.

Beven, K. 2019a How to make advances in hydrological modelling. *Hydrology Research.* https://doi.org/10.2166/nh.2019.134

Beven, K. 2019b Towards a methodology for testing models as hypotheses in the inexact sciences. *Proceedings of the Royal Society A* **475**, 20180862.

Beven, K., Cloke, H., Pappenberger, F., Lamb, R. & Hunter, N. 2015 Hyperresolution information and hyperresolution ignorance in modelling the hydrology of the land surface. *Science China Earth Sciences* **58**, 25–35.

Bierkens, M. F. 2015 Global hydrology 2015: state, trends, and directions. *Water Resources Research* **51**, 4923–4947.

Blöschl, G., Bierkens, M. F., Chambel, A., Cudennec, C., Destouni, G., Fiori, A., Kirchner, J. W., McDonnell, J. J., Savenije, H. H., Sivapalan, M. & Stumpp, C. 2019 Twenty-three unsolved problems in hydrology (UPH)–a community perspective. *Hydrological Sciences Journal* **64**, 1141–1158.

British Geological Survey (BGS). 2019 *Gateway to the Earth: Science Strategy for the British Geological Survey 2019–2023.* British Geological Survey, Keyworth.

Broderick, C., Murphy, C., Wilby, R. L., Matthews, T., Prudhomme, C. & Adamson, M. 2019 National assessment of climate change allowances for future flood risk using a scenario-neutral framework. *Water Resources Research* **55**, 1079–1104.

Bronstert, A., Agarwal, A., Boessenkool, B., Crisologo, I., Fischer, M., Heistermann, M., Köhn-Reich, L., López-Tarazón, J. A., Moran, T., Ozturk, U. & Reinhardt-Imjela, C. 2018 Forensic hydro-meteorological analysis of an extreme flash flood: the 2016-05-29 event in Braunsbach, SW Germany. *Science of the Total Environment* **630**, 977–991.

Burt, S. & Hawkins, E. 2019 Near-zero humidities on Ben Nevis, Scotland, revealed by pioneering nineteenth-century observers and modern volunteers. *International Journal of Climatologyt* **39**, 4451–4466. https://doi.org/10.1002/joc.6084

Carpenter, S. R., Mooney, H. A., Agard, J., Capistrano, D., DeFries, R. S., Díaz, S., Dietz, T., Duraiappah, A. K., Oteng-Yeboah, A., Pereira, H. M. & Perrings, C. 2009 Science for managing ecosystem services: beyond the millennium ecosystem assessment. *Proceedings of the National Academy of Sciences* **106**, 1305–1312.

Chakraborti, D., Mukherjee, S. C., Pati, S., Sengupta, M. K., Rahman, M. M., Chowdhury, U. K., Lodh, D., Chanda, C. R., Chakraborti, A. K. & Basu, G. K. 2003 Arsenic groundwater contamination in Middle Ganga Plain, Bihar, India: a future danger? *Environmental Health Perspectives* **111**, 1194–1201.

Chen, Y. & Han, D. 2016 Big data and hydroinformatics. *Journal of Hydroinformatics* **18**, 599–614.

Climate Outreach 2016 *Communicating Drought Risk in a Changing Climate*. Climate Outreach, Oxford.

Coles, D., Yu, D., Wilby, R. L., Green, D. & Herring, Z. 2017 Beyond 'flood hotspots': modelling emergency service accessibility during flooding in York, UK. *Journal of Hydrology* **546**, 419–436.

Collet, L., Harrigan, S., Prudhomme, C., Formetta, G. & Beevers, L. 2018 Future hot-spots for hydro-hazards in Great Britain: a probabilistic assessment. *Hydrology and Earth System Sciences* **22**, 5387–5401.

Conway, D., Van Garderen, E. A., Deryng, D., Dorling, S., Krueger, T., Landman, W., Lankford, B., Lebek, K., Osborn, T., Ringler, C. & Thurlow, J. 2015 Climate and Southern Africa's water–energy–food nexus. *Nature Climate Change* **5**, 837–846.

Conway, D., Nicholls, R. J., Brown, S., Tebboth, M. G., Adger, W. N., Ahmad, B., Biemans, H., Crick, F., Lutz, A. F., De Campos, R. S. & Said, M. 2019 The need for bottom-up assessments of climate risks and adaptation in climate-sensitive regions. *Nature Climate Change* **9**, 503–511.

Cook, E. R., Seager, R., Kushnir, Y., Briffa, K. R., Büntgen, U., Frank, D., Krusic, P. J., Tegel, W., van der Schrier, G., Andreu-Hayles, L. & Baillie, M. 2015 Old world megadroughts and pluvials during the Common Era. *Science Advances* **10**, e1500561.

Courty, L. G., Wilby, R. L., Hillier, J. & Slater, L. J. 2019 Intensity-duration-frequency curves of precipitation at the global scale. *Environmental Research Letters* **14**, 084045.

Cutter, S. L. 2018 Compound, cascading, or complex disasters: what's in a name? *Environment* **60**, 16–25.

d'Amour, C. B., Wenz, L., Kalkuhl, M., Steckel, J. C. & Creutzig, F. 2016 Teleconnected food supply shocks. *Environmental Research Letters* **11**, 035007.

De Luca, P., Hillier, J. K., Wilby, R. L., Quinn, N. W. & Harrigan, S. 2017 Extreme multi-basin flooding linked with extra-tropical cyclones. *Environmental Research Letters* **12**, 114009.

de Magny, G. C., Thiaw, W., Kumar, V., Manga, N. M., Diop, B. M., Gueye, L., Kamara, M., Roche, B., Murtugudde, R. & Colwell, R. R. 2012 Cholera outbreak in Senegal in 2005: was climate a factor? *PLoS One* **7**, e44577.

Demeritt, D. & Nobert, S. 2014 Models of best practice in flood risk communication and management. *Environmental Hazards* **13**, 313–328.

Di Baldassarre, G., Viglione, A., Carr, G., Kuil, L., Yan, K., Brandimarte, L. & Blöschl, G. 2015 Debates – perspectives on socio-hydrology: capturing feedbacks between physical and social processes. *Water Resources Research* **51**, 4770–4781.

Dixon, S. G. & Wilby, R. L. 2016 Forecasting reservoir inflows using remotely sensed precipitation estimates: a pilot study for the River Naryn, Kyrgyzstan. *Hydrological Sciences Journal* **61**, 107–122.

Dixon, S. G. & Wilby, R. L. 2019 A seasonal forecasting procedure for reservoir inflows in Central Asia. *Rivers Research and Applications – Special Issue*. doi:10.1002/rra.3506.

Dobler, C., Hagemann, S., Wilby, R. L. & Stätter, J. 2012 Quantifying different sources of uncertainty in hydrological projections in an Alpine watershed. *Hydrology and Earth System Sciences* **16**, 4343–4360.

Douglas, I., Alam, K., Maghenda, M., Mcdonnell, Y., McLean, L. & Campbell, J. 2008 Unjust waters: climate change, flooding and the urban poor in Africa. *Environment and Urbanization* **20**, 87–205.

Ekström, M., Gutmann, E. D., Wilby, R. L., Tye, M. R. & Kirono, D. G. C. 2018 Robustness of hydroclimate metrics for climate change impact research. *WIRES Water*. doi.org/10.1002/wat2.1288.

Emerton, R., Cloke, H. L., Stephens, E. M., Zsoter, E., Woolnough, S. J. & Pappenberger, F. 2017 Complex picture for likelihood of ENSO-driven flood hazard. *Nature Communications* **8**, 14796.

Environment Agency 2016 *Adapting to Climate Change: Advice for Flood and Coastal Erosion Risk Management Authorities*. Environment Agency, Bristol. Available from: https://www.gov.uk/government/publications/adapting-to-climate-change-for-risk-management-authorities (accessed 26 June 2019).

Environment Agency and Natural Resource Wales 2018 *Water Resource Planning Guideline: Interim Update*. July 2018, Bristol.

Falkenmark, M. 2013 Growing water scarcity in agriculture: future challenge to global water security. *Philosophical*

Transactions of the Royal Society A: Mathematical, Physical and Engineering Sciences **371**, 20120410.

Fang, S., Xu, L., Zhu, Y., Liu, Y., Liu, Z., Pei, H., Yan, J. & Zhang, H. 2015 An integrated information system for snowmelt flood early-warning based on internet of things. *Information Systems Frontiers* **17**, 321–335.

Folland, C. K., Hannaford, J., Bloomfield, J. P., Kendon, M., Svensson, C., Marchant, B. P., Prior, J. & Wallace, E. 2015 Multi-annual droughts in the English lowlands: a review of their characteristics and climate drivers in the winter half-year. *Hydrology and Earth System Sciences* **19**, 2353–2375.

Fowler, H. J. & Wilby, R. L. 2010 Detecting changes in seasonal precipitation extremes using regional climate model projections: implications for managing fluvial flood risk. *Water Resources Research* **46**, W03525.

Gao, Y., Chen, F., Lettenmaier, D. P., Xu, J., Xiao, L. & Li, X. 2018 Does elevation-dependent warming hold true above 5000 m elevation? Lessons from the Tibetan plateau. *npj Climate and Atmospheric Science* **1**, Article number: 19.

Gell, P. A., Reid, M. A. & Wilby, R. L. 2019 Management pathways for the floodplain wetlands of the southern Murray darling basin: lessons from history. *Rivers Research and Applications – Special Issue.* doi:10.1002/rra.3515.

Gill, J. C. & Malamud, B. D. 2017 Anthropogenic processes, natural hazards, and interactions in a multi-hazard framework. *Earth-Science Reviews* **166**, 246–269.

Gleeson, T., Wada, Y., Bierkens, M. F. & van Beek, L. P. 2012 Water balance of global aquifers revealed by groundwater footprint. *Nature* **488**, 197–200.

Gleick, P. H. & Palaniappan, M. 2010 Peak water limits to freshwater withdrawal and use. *Proceedings of the National Academy of Sciences* **107**, 11155–11162.

Gober, P. & Wheater, H. S. 2015 Debates – perspectives on socio-hydrology: modeling flood risk as a public policy problem. *Water Resources Research* **51**, 4782–4788.

Gough, K. V., Yankson, P. W. K., Wilby, R. L., Amankwaa, E., Abarike, M., Codjoe, S. N. A., Griffiths, P. L., Kasei, R., Kayaga, S. & Nabilse, C. K. 2019 Vulnerability to extreme weather events in cities: implications for infrastructure and livelihoods. *Journal of the British Academy* **7**, 155–181.

Gouweleeuw, B. T., Kvas, A., Gruber, C., Gain, A. K., Mayer-Gürr, T., Flechtner, F. & Güntner, A. 2018 Daily GRACE gravity field solutions track major flood events in the Ganges–Brahmaputra Delta. *Hydrology and Earth System Sciences* **22**, 2867–2880.

Green, J. M., Cranston, G. R., Sutherland, W. J., Tranter, H. R., Bell, S. J., Benton, T. G., Blixt, E., Bowe, C., Broadley, S., Brown, A. & Brown, C. 2017a Research priorities for managing the impacts and dependencies of business upon food, energy, water and the environment. *Sustainability Science* **12**, 319–331.

Green, D., Yu, D., Pattison, I., Wilby, R. L., Bosher, L., Patel, R., Thompson, P., Trowell, K., Draycon, J., Halse, M. & Yang, L. 2017b City-scale accessibility of emergency responders operating during flood events. *Natural Hazards and Earth System Sciences* **17**, 1–16.

Gunawardena, K. R., Wells, M. J. & Kershaw, T. 2017 Utilising green and bluespace to mitigate urban heat island intensity. *Science of the Total Environment* **584**, 1040–1055.

Haasnoot, M., Kwakkel, J. H., Walker, W. E. & ter Maat, J. 2013 Dynamic adaptive policy pathways: a method for crafting robust decisions for a deeply uncertain world. *Global Environmental Change* **23**, 485–498.

Hallegatte, S. 2009 Strategies to adapt to an uncertain climate change. *Global Environmental Change* **19**, 240–247.

Hallegatte, S., Green, C., Nicholls, R. J. & Corfee-Morlot, J. 2013 Future flood losses in major coastal cities. *Nature Climate Change* **3**, 802–806.

Hannah, D. M., Brown, L. E., Milner, A. M., Gurnell, A. M., McGregor, G. R., Petts, G. E., Smith, B. P. & Snook, D. L. 2007 Integrating climate–hydrology–ecology for alpine river systems. *Aquatic Conservation: Marine and Freshwater Ecosystems* **17**, 636–656.

Harmancioglu, N. B. 2017 Overview of water policy developments: pre-and post-2015 development agenda. *Water Resources Management* **31**, 3001–3021.

Harrigan, S., Murphy, C., Hall, J., Wilby, R. L. & Sweeney, J. 2014 Attribution of detected changes in streamflow using multiple working hypotheses. *Hydrology and Earth System Sciences* **18**, 1935–1952.

Henonin, J., Russo, B., Mark, O. & Gourbesville, P. 2013 Real-time urban flood forecasting and modelling–a state of the art. *Journal of Hydroinformatics* **15**, 717–736.

Herath, S. M., Sarukkalige, P. R. & Nguyen, V. T. V. 2016 A spatial temporal downscaling approach to development of IDF relations for Perth airport region in the context of climate change. *Hydrological Sciences Journal* **61**, 2061–2070.

Hillier, J. K., Macdonald, N., Leckebusch, G. C. & Stavrinides, A. 2015 Interactions between apparently 'primary' weather-driven hazards and their cost. *Environmental Research Letters* **10**, 104003.

Hirabayashi, Y., Mahendran, R., Koirala, S., Konoshima, L., Yamazaki, D., Watanabe, S., Kim, H. & Kanae, S. 2013 Global flood risk under climate change. *Nature Climate Change* **3**, 816–821.

Hunt, A. S. P., Wilby, R. L., Dale, N., Sura, K. & Watkiss, P. 2014 Embodied water imports to the UK under climate change. *Climate Research* **59**, 89–101.

Huss, M. & Hock, R. 2018 Global-scale hydrological response to future glacier mass loss. *Nature Climate Change* **8**, 135–140.

Immerzeel, W. W., Van Beek, L. P. & Bierkens, M. F. 2010 Climate change will affect the Asian water towers. *Science* **328**, 1382–1385.

International Hydropower Association 2019 *Hydropower Sector Climate Resilience Guide.* International Hydropower Association, London. Available from: https://www.hydropower.org/publications/hydropower-sector-climate-resilience-guide (accessed 26 June 2019).

Ionita, M., Lohmann, G., Rimbu, N., Chelcea, S. & Dima, M. 2012 Interannual to decadal summer drought variability over Europe and its relationship to global sea surface temperature. *Climate Dynamics* **38**, 363–377.

IPCC 2018 Summary for Policymakers. In: *Global warming of 1.5 °C. An IPCC Special Report on the impacts of global warming of 1.5 °C above pre-industrial levels and related global greenhouse gas emission pathways, in the context of strengthening the global response to the threat of climate change, sustainable development, and efforts to eradicate poverty*. Available from: https://www.ipcc.ch/site/assets/uploads/sites/2/2018/07/SR15_SPM_High_Res.pdf (accessed 24 June 2019).

Jennings, B., Heltne, P. & Kintzele, K. 2009 Principles of water ethics. *Minding Nature* August 2009, 25–28.

Katsman, C. A., Sterl, A., Beersma, J. J., Van den Brink, H. W., Church, J. A., Hazeleger, W., Kopp, R. E., Kroon, D., Kwadijk, J., Lammersen, R. & Lowe, J. 2011 Exploring high-end scenarios for local sea level rise to develop flood protection strategies for a low-lying delta – the Netherlands as an example. *Climatic Change* **109**, 617–645.

KPMG 2014 *Future State 2030: The global megatrends shaping governments*. Available from: https://assets.kpmg/content/dam/kpmg/pdf/2014/02/future-state-2030-v3.pdf (accessed 24 June 2019).

Krause, S., Hannah, D. M., Fleckenstein, J. H., Heppell, C. M., Kaeser, D., Pickup, R., Pinay, G., Robertson, A. L. & Wood, P. J. 2011 Inter-disciplinary perspectives on processes in the hyporheic zone. *Ecohydrology* **4**, 481–499.

Kundzewicz, Z. W. & Stakhiv, E. Z. 2010 Are climate models 'ready for prime time' in water resources management applications, or is more research needed? *Hydrological Sciences Journal* **55**, 1085–1089.

Kure, S., Jang, S., Ohara, N., Kavvas, M. L. & Chen, Z. Q. 2013 Hydrologic impact of regional climate change for the snowfed and glacierfed river basins in the Republic of Tajikistan: hydrological response of flow to climate change. *Hydrological Processes* **27**, 4057–4070.

Leedal, D., Neal, J., Beven, K., Young, P. & Bates, P. 2010 Visualization approaches for communicating real-time flood forecasting level and inundation information. *Journal of Flood Risk Management* **3**, 140–150.

Liepert, B. G. & Lo, F. 2013 CMIP5 update of 'Inter-model variability and biases of the global water cycle in CMIP3 coupled climate models'. *Environmental Research Letters* **8**, 029401.

Liepert, B. G. & Previdi, M. 2012 Inter-model variability and biases of the global water cycle in CMIP3 coupled climate models. *Environmental Research Letters* **7**, 014006.

Luo, Y., Wang, X., Piao, S., Sun, L., Ciais, P., Zhang, Y., Ma, C., Gan, R. & He, C. 2018 Contrasting streamflow regimes induced by melting glaciers across the Tien Shan–Pamir–North Karakoram. *Scientific Reports* **8**, 16470.

Lutz, A., Immerzeel, W. W., Gobiet, A., Pellicciotti, F. & Bierkens, M. F. P. 2013 Comparison of climate change signals in CMIP3 and CMIP5 multi-model ensembles and implications for Central Asian glaciers. *Hydrology and Earth System Sciences* **17**, 3661–3677.

Lutz, A. F., Immerzeel, W. W., Shrestha, A. B. & Bierkens, M. F. P. 2014 Consistent increase in High Asia's runoff due to increasing glacier melt and precipitation. *Nature Climate Change* **4**, 587–592.

Matthews, T., Wilby, R. L. & Murphy, C. 2019 An emerging tropical cyclone-deadly heat compound hazard. *Nature Climate Change* **9**, 602–606.

Maurer, J. M., Schaefer, J. M., Rupper, S. & Corley, A. 2019 Acceleration of ice loss across the Himalayas over the past 40 years. *Science Advances* **5**, eaav7266.

McCabe, M. F., Rodell, M., Alsdorf, D. E., Miralles, D. G., Uijlenhoet, R., Wagner, W., Lucieer, A., Houborg, R., Verhoest, N. E., Franz, T. E. & Shi, J. 2017 The future of Earth observation in hydrology. *Hydrology and Earth System Sciences* **21**, 3879–3914.

McGregor, G. R. 2019 Climate and rivers. *Rivers Research and Applications – Special Issue*. doi:10.1002/rra.3508.

McSweeney, C. F., Jones, R. G., Lee, R. W. & Rowell, D. P. 2015 Selecting CMIP5 GCMs for downscaling over multiple regions. *Climate Dynamics* **44**, 3237–3260.

Merz, B., Vorogushyn, S., Uhlemann, S., Delgado, J. & Hundecha, Y. 2012 HESS opinions more efforts and scientific rigour are needed to attribute trends in flood time series'. *Hydrology and Earth System Sciences* **16**, 1379–1387.

Montanari, A., Young, G., Savenije, H. H. G., Hughes, D., Wagener, T., Ren, L. L., Koutsoyiannis, D., Cudennec, C., Toth, E., Grimaldi, S. & Blöschl, G. 2013 'Panta Rhei – everything flows': change in hydrology and society – the IAHS scientific decade 2013–2022. *Hydrological Sciences Journal* **58**, 1256–1275.

Murphy, C., Wilby, R. L., Matthews, T. K., Thorne, P., Broderick, C., Fealy, R., Hall, J., Harrigan, S., Jones, P., McCarthy, G. & Macdonald, N. 2019 Wetter winters and drier summers in England and Wales precipitation explained by observational and sampling bias in early records. *International Journal of Climatology*. doi:10.1002/joc.6208.

Muthusamy, M., Rivas Casado, M., Salmoral, G., Irvine, T. & Leinster, P. 2019 A remote sensing based integrated approach to quantify the impact of fluvial and pluvial flooding in an urban catchment. *Remote Sensing* **11**, 577.

Najmaddin, P. M., Whelan, M. J. & Balzter, H. 2017 Estimating daily reference evapotranspiration in a semi-arid region using remote sensing data. *Remote Sensing* **9**, 779.

National Academies of Sciences, Engineering, and Medicine. 2018 *Future Water Priorities for the Nation: Directions for the U.S. Geological Survey Water Mission Area*. The National Academies Press, Washington, DC.

Naumann, G., Alfieri, L., Wyser, K., Mentaschi, L., Betts, R. A., Carrao, H., Spinoni, J., Vogt, J. & Feyen, L. 2018 Global changes in drought conditions under different levels of warming. *Geophysical Research Letters* **45**, 3285–3296.

Netherlands Ministry of Infrastructure and Environment 2014 *Guideline for Stress Testing the Climate Resilience of Urban*

Areas. Available from: https://climate-adapt.eea.europa.eu/metadata/guidances/guideline-for-stress-testing-the-climate-resilience-of-urban-areas/11258895 (accessed 26 June 2019).

New York City Panel on Climate Change 2013 Climate Risk Information 2013: Observations, Climate Change Projections, and Maps. In: *NPCC2* (C. Rosenzweig & W. Solecki, eds). Prepared for use by the City of New York Special Initiative on Rebuilding and Resiliancy, Mayor's Office of Long-Term Planning & Sustainability, New York, p. 38.

Ohara, N., Jang, S., Kure, S., Richard Chen, Z. Q. & Kavvas, M. L. 2014 Modeling of interannual snow and ice storage in high-altitude regions by dynamic equilibrium concept. *Journal of Hydrologic Engineering* **19**, 04014034.

Oreskes, N., Shrader-Frechette, K. & Belitz, K. 1994 Verification, validation, and confirmation of numerical models in the earth sciences. *Science* **263**, 641–646.

Pahl-Wostl, C., Voeroesmarty, C., Bhaduri, A., Bogardi, J., Rockstroem, J. & Alcamo, J. 2013 Towards a sustainable water future: shaping the next decade of global water research. *Current Opinion in Environmental Sustainability* **5**, 708–714.

Pappenberger, F. & Beven, K. J. 2006 Ignorance is bliss: or seven reasons not to use uncertainty analysis. *Water Resources Research* **42**, W05302.

Pepin, N., Bradley, R. S., Diaz, H. F., Baraër, M., Caceres, E. B., Forsythe, N., Fowler, H., Greenwood, G., Hashmi, M. Z., Liu, X. D. & Miller, J. R. 2015 Elevation-dependent warming in mountain regions of the world. *Nature Climate Change* **5**, 424–430.

Pielke Sr., R. A., Wilby, R., Niyogi, D., Hossain, F., Dairuku, K., Adegoke, J., Kallos, G., Seastedt, T. & Suding, K. 2012 Dealing with complexity and extreme events using a bottom-up, resource-based vulnerability perspective. *Extreme Events and Natural Hazards: The Complexity Perspective, Geophysical Monograph Series* **196**, 345–359.

Pritchard, D. M., Forsythe, N., Fowler, H. J., O'Donnell, G. M. & Li, X. F. 2019 Evaluation of Upper Indus near-surface climate representation by WRF in the high Asia refined analysis. *Journal of Hydrometeorology* **20**, 467–487.

Prudhomme, C., Giuntoli, I., Robinson, E. L., Clark, D. B., Arnell, N. W., Dankers, R., Fekete, B. M., Franssen, W., Gerten, D., Gosling, S. N. & Hagemann, S. 2014 Hydrological droughts in the 21st century, hotspots and uncertainties from a global multi model ensemble experiment. *Proceedings of the National Academy of Sciences of the United States of America* **111**, 3262–3267.

Pulwarty, R. S. & Sivakumar, M. V. 2014 Information systems in a changing climate: early warnings and drought risk management. *Weather and Climate Extremes* **3**, 14–21.

Qiuming, W., Xiyun, J., Yu, P. & Shengrong, H. 2012 Intelligent micro-irrigation system based on internet of things in arid area. *Transactions of the Chinese Society of Agricultural Engineering* **2012** (1).

Ranger, N., Reeder, T. & Lowe, J. 2013 Addressing 'deep' uncertainty over long-term climate in major infrastructure projects: four innovations of the Thames Estuary 2100 Project. *EURO Journal on Decision Processes* **1**, 233–262.

Reid, A .J., Carlson, A. K., Creed, I. F., Eliason, E. J., Gell, P. A., Johnson, P. T., Kidd, K. A., MacCormack, T. J., Olden, J. D., Ormerod, S. J. & Smol, J. P. 2019 Emerging threats and persistent conservation challenges for freshwater biodiversity. *Biological Reviews* **94**, 849–873.

Riley, W. D., Potter, E. C., Biggs, J., Collins, A. L., Jarvie, H. P., Jones, J. I., Kelly-Quinn, M., Ormerod, S. J., Sear, D. A., Wilby, R. L. & Broadmeadow, S. 2018 Small water bodies in Great Britain and Ireland: ecosystem function, human-generated degradation, options for restorative action. *Science of the Total Environment* **645**, 1598–1616.

Rodell, M., Famiglietti, J. S., Wiese, D. N., Reager, J. T., Beaudoing, H. K., Landerer, F. W. & Lo, M. H. 2018 Emerging trends in global freshwater availability. *Nature* **557**, 651–659.

Saft, M., Western, A. W., Zhang, L., Peel, M. C. & Potter, N. J. 2015 The influence of multiyear drought on the annual rainfall-runoff relationship: an Australian perspective. *Water Resources Research* **51**, 2444–2463.

Saft, M., Peel, M. C., Western, A. W., Perraud, J. M. & Zhang, L. 2016 Bias in streamflow projections due to climate-induced shifts in catchment response. *Geophysical Research Letters* **43**, 1574–1581.

Samaniego, L., Kumar, R. & Jackisch, C. 2011 Predictions in a data-sparse region using a regionalized grid-based hydrologic model driven by remotely sensed data. *Hydrology Research* **42**, 338–355.

Serinaldi, F. & Kilsby, C. G. 2015 Stationarity is undead: uncertainty dominates the distribution of extremes. *Advances in Water Resources* **77**, 17–36.

Shannon, S., Smith, R., Wiltshire, A., Payne, T., Huss, M., Betts, R., Caesar, J., Koutroulis, A., Jones, D. & Harrison, S. 2019 Global glacier volume projections under high-end climate change scenarios. *Cryosphere* **13**, 325–350.

Shaposhnikov, D., Revich, B., Bellander, T., Bedada, G. B., Bottai, M., Kharkova, T., Kvasha, E., Lezina, E., Lind, T., Semutnikova, E. & Pershagen, G. 2014 Mortality related to air pollution with the Moscow heat wave and wildfire of 2010. *Epidemiology* **25**, 359–364.

Siegfried, T., Bernauer, T., Guiennet, R., Sellars, S., Robertson, A. W., Mankin, J., Bauer-Gottwein, P. & Yakovlev, A. 2012 Will climate change exacerbate water stress in Central Asia? *Climatic Change* **112**, 881–899.

Sivapalan, M., Savenije, H. H. & Blöschl, G. 2012 Socio-hydrology: a new science of people and water. *Hydrological Processes* **26**, 1270–1276.

Slater, L. J., Khouakhi, A. & Wilby, R. L. 2019 River channel conveyance capacity adjusts to modes of climate variability. *Scientific Reports* **9**, 12619.

Sorg, A., Huss, M., Rohrer, M. & Stoffel, M. 2014 The days of plenty might soon be over in glacierized Central Asian catchments. *Environmental Research Letters* **9**, 104018.

Sun, Q., Miao, C., Duan, Q., Ashouri, H., Sorooshian, S. & Hsu, K. L. 2018 A review of global precipitation data sets: data sources, estimation, and intercomparisons. *Reviews of Geophysics* **56**, 79–107.

Suresh, S. 2012 Research funding: global challenges need global solutions. *Nature* **490**, 337–338.

Sutton, R. T. & Dong, B. 2012 Atlantic Ocean influence on a shift in European climate in the 1990s. *Nature Geoscience* **5**, 788–792.

United States Army Corps of Engineers 2013 *Incorporating Sea Level Change in Civil Works ER 1110-2-8162: Sea Level Change Calculator.*

van Dijk, A. I. J. M., Beck, H. E., Crosbie, R. S., de Jeu, R. A., Liu, Y. Y., Podger, G. M., Timbal, B. & Viney, N. R. 2013 The Millennium Drought in southeast Australia (2001–2009): natural and human causes and implications for water resources, ecosystems, economy, and society. *Water Resources Research* **49**, 1040–1057.

Verdon-Kidd, D. C. & Kiem, A. S. 2009 Nature and causes of protracted droughts in southeast Australia: comparison between the Federation, WWII, and Big Dry droughts. *Geophysical Research Letters* **36**, L22707.

Visser-Quinn, A., Beevers, L., Collet, L., Formetta, G., Smith, K., Wanders, N., Thober, S., Pan, M. & Kumar, R. 2019 Spatio-temporal analysis of compound hydro-hazard extremes across the UK. *Advances in Water Resources* **130**, 77–90.

Vogel, M. M., Zscheischler, J. & Seneviratne, S. I. 2018 Varying soil moisture–atmosphere feedbacks explain divergent temperature extremes and precipitation projections in central Europe. *Earth System Dynamics* **9**, 1107–1125.

Vörösmarty, C. J., McIntryre, P. B. & Gessner, M. O. 2010 Global threats to human water security and river biodiversity. *Nature* **467**, 555–561.

Vugteveen, P., Lenders, R. & Van Den Besselaar, P. 2014 The dynamics of interdisciplinary research fields: the case of river research. *Scientometrics* **100**, 73–96.

Wagener, T., Sivapalan, M., Troch, P. A., McGlynn, B. L., Harman, C. J., Gupta, H. V., Kumar, P., Rao, P. S. C., Basu, N. B. & Wilson, J. S. 2010 The future of hydrology: an evolving science for a changing world. *Water Resources Research* **46**, W05301.

Wahl, T., Jain, S., Bender, J., Meyers, S. D. & Luther, M. E. 2015 Increasing risk of compound flooding from storm surge and rainfall for major US cities. *Nature Climate Change* **5**, 1093–1097.

Walsh, C. J., Roy, A. H., Feminella, J. W., Cottingham, P. D., Groffman, P. M. & Morgan, R. P. 2005 The urban stream syndrome: current knowledge and the search for a cure. *Journal of the North American Benthological Society* **24**, 706–723.

Watts, G. 2015 Hydrology with impact: how does hydrological science inform decision-makers? *Hydrology Research* **47**, 545–551.

Wheater, H. S. & Gober, P. 2015 Water security and the science agenda. *Water Resources Research* **51**, 5406–5424.

Wilby, R. L. 2006 When and where might climate change be detectable in UK river flows? *Geophysical Research Letters* **33**, L19407. doi:10.1029/2006GL027552.

Wilby, R. L. 2017 *How Connected is Climate Change to Other Global Challenges? Climate Change in Practice*, Chapter 18. Cambridge University Press, Cambridge.

Wilby, R. L. & Keenan, R. 2012 Adapting to flood risk under climate change. *Progress in Physical Geography* **36**, 349–379.

Wilby, R. L. & Yu, D. 2013 Rainfall and temperature estimation for a data sparse region. *Hydrology and Earth System Sciences* **17**, 3937–3955.

Wilby, R. L., Prudhomme, C., Parry, S. & Muchan, K. G. L. 2015 Persistence of hydrometeorological droughts in the United Kingdom: a regional analysis of multi-season rainfall and river flow anomalies. *Journal of Extreme Events* **2**, 1550006.

Wilby, R. L., Clifford, N. J., De Luca, P., Harrigan, S., Hillier, J. K., Hodgkins, R., Johnson, M. F., Matthews, T. K., Murphy, C., Noone, S. J. & Parry, S. 2017 The 'dirty dozen' of freshwater science: detecting then reconciling hydrological data biases and errors. *WIRES Water* **4**, e1209. doi:10.1002/wat2.1209.

Wilkinson, F. 2019 Inside the Everest expedition that built the world's highest weather station. *National Geographic.* Available from: https://www.nationalgeographic.com/adventure/2019/06/mount-everest-highest-weather-station/ (accessed 24 June 2019).

Wimmer, F., Schlaffer, S. & Menzel, L. 2009 Distributed modelling of climate change impacts on snow sublimation in Northern Mongolia. *Advances in Geosciences* **21**, 117–124.

Wood, E. F., Roundy, J. K., Troy, T. J., Van Beek, L. P. H., Bierkens, M. F., Blyth, E., de Roo, A., Döll, P., Ek, M., Famiglietti, J. & Gochis, D. 2011 Hyperresolution global land surface modeling: meeting a grand challenge for monitoring Earth's terrestrial water. *Water Resources Research* **47**, W05301.

Woodward, G., Gessner, M. O., Giller, P. S., Gulis, V., Hladyz, S., Lecerf, A., Malmqvist, B., McKie, B. G., Tiegs, S. D., Cariss, H. & Dobson, M. 2012 Continental-scale effects of nutrient pollution on stream ecosystem functioning. *Science* **336**, 1438–1440.

Xu, L., Gober, P., Wheater, H. S. & Kajikawa, Y. 2018 Reframing socio-hydrological research to include a social science perspective. *Journal of Hydrology* **563**, 76–83.

Yang, Y. E., Wi, S., Ray, P. A., Brown, C. M. & Khalil, A. F. 2016 The future nexus of the Brahmaputra river basin: climate, water, energy and food trajectories. *Global Environmental Change* **37**, 16–30.

Yin, J., Yu, D. & Wilby, R. L. 2015 Modelling the impact of land subsidence on urban pluvial flooding: a case study of downtown Shanghai, China. *Science of the Total Environment* **544**, 744–753.

Yu, D., Yin, J., Wilby, R. L., Lane, S., Aerts, J., Lin, N., Liu, M., Yuan, H., Chen, J., Prudhomme, C., Guan, M., Baruch, A., Johnson, W., Tang, X., Yu, L. & Xu, S. 2019 Low magnitude floods dramatically impact emergency response for vulnerable populations. *Nature Sustainability*. Under review.

Zhang, D., Lindholm, G. & Ratnaweera, H. 2018 Use long short-term memory to enhance internet of things for combined sewer overflow monitoring. *Journal of Hydrology* **556**, 409–418.

Ziegler, A. D., Maurer, E. P., Sheffield, J., Nijssen, B., Wood, E. F. & Lettenmaier, D. P. 2005 Detection time for plausible changes in annual precipitation, evapotranspiration, and streamflow in three Mississippi River sub-basins. *Climatic Change* **72**, 17–36.

Zscheischler, J., Westra, S., Van Den Hurk, B. J., Seneviratne, S. I., Ward, P. J., Pitman, A., AghaKouchak, A., Bresch, D. N., Leonard, M., Wahl, T. & Zhang, X. 2018 Future climate risk from compound events. *Nature Climate Change* **8**, 469–477.

Žuvela-Aloise, M., Koch, R., Buchholz, S. & Früh, B. 2016 Modelling the potential of green and blue infrastructure to reduce urban heat load in the city of Vienna. *Climatic Change* **135**, 425–438.

First received 1 July 2019; accepted in revised form 9 October 2019. Available online 25 October 2019

How to make advances in hydrological modelling

Keith Beven

ABSTRACT

After some background about what I have learned from a career in hydrological modelling, I present some opinions about how we might make progress in improving hydrological models in future, including how to decide whether a model is fit for purpose; how to improve process representations in hydrological models; and how to take advantage of Models of Everywhere. Underlying all those issues, however, is the fundamental problem of improving the hydrological data available for both forcing and evaluating hydrological models. It would be a major advance if the hydrological community could come together to prioritise and commission the new observational methods that are required to make real progress.

Key words | epistemic uncertainties, hypothesis testing, model invalidation, Models of Everywhere, process representations

Keith Beven†
Lancaster Environment Centre,
Lancaster University,
Lancaster,
UK
E-mail: *k.beven@lancaster.ac.uk*

†President's Invited Address, BHS Symposium, University of Westminster, September 2018.

SOME BACKGROUND

My first attempt at a hydrological model was produced as an undergraduate student at the University of Bristol in about 1970. It was an attempt to model the famous Lynmouth Flood in 1952. It was programmed in Algol and physically existed as a pack of punched cards that needed to be fed into a card reader every time a run was made (compilation errors, run-time errors, and eventually, production runs included). The primary data available were rainfall records, so the only 'calibration' data were indirect post-flood estimates of a peak discharge. This was a high sediment laden flow that transported some huge boulders, so any such estimate would have been highly uncertain. Even so, that simple study taught me a great deal about the importance of antecedent conditions in trying to predict flood discharges; the wetting of the catchment prior to the flood was extremely important (as has also been the case in many more recent cases of flash flooding in the UK).

In starting my PhD at the University of East Anglia in Norwich in 1971, I made a survey of hydrological models in the literature. Even at that time, there were a plethora of different models. With the more widespread availability of digital computers in the late 1960s, many PhD projects and consultants were producing their own models. Most of these were conceptual models of the Stanford Watershed Model type, which itself was the PhD project of Norman Crawford at Standford University under the direction of Ray K. Linsley (Crawford & Linsley 1966). This model was the foundation of the Hydrocomp consultancy, and I met both of them when I was able to participate, while still a PhD student myself, at the first UK Hydrocomp workshop. The Hydrocomp Simulation Programme in Fortran (HSPF) was later adopted by the US EPA and remained in use as a freely available tool. When I gave up my count of models at over 100 in 1971, I was already asking the question of how can we do better?

My response was to try and be objective, to base a model on best physical principles, and to measure rather than calibrate the model parameters. Al Freeze was already advocating this in the Freeze and Harlan paper in *Journal of Hydrology* in 1969 (Freeze & Harlan 1969) and

doi: 10.2166/nh.2019.134

implementing it using finite difference methods at the Thomas J. Watson research centre of IBM at Yorktown Heights. I took a slightly different strategy using finite element methods to solve the Richards equation so that my hillslopes and soil horizons on those hillslopes could look more natural in the discretisation grid (I had a physical geography rather than engineering degree after all!). I did not have quite the same resources as Al Freeze. The model was implemented as two full boxes of computer cards (with all the same issues of compilation and run-time errors, but now with many more cards to get through the card reader successfully) and ran on an ICL1904 mainframe computer. I also carried out the laboratory work necessary to determine all the soil moisture characteristics on soil cores and the fieldwork necessary for channel cross-sections and roughness. The model was applied to the East Twin catchment in the Mendips that had been studied by Darrell Weyman (1970, 1973) for his PhD, and the results were really rather bad, a fact noted by my PhD examiner, Terrence O'Donnell. They were finally published as part of my Dalton Lecture paper (Beven 2001) which tells the story of how that experience shaped my research career.

I was fortunate to then work with Mike Kirkby as a post-doc at the University of Leeds on the development of Topmodel (see Beven & Kirkby 1979) based on Mike's concept of the topographic index. Given my experience of physically based modelling, I was more than happy to take another approach, but still one that allowed the results of the modelling to be mapped back into space (I was again in a Geography Department). The only problem with that was that both the topographic analysis that went into Topmodel and the analysis of the spatial nature of the results had to be done manually. There were no Digital Terrain Models, and computer outputs were still on lineprinter paper. We were also running a nested catchment experiment, with both rainfall and stream-level data recorded on paper charts so a great deal of time was spent just getting the data into a computer-compatible form (e.g. Beven & Callen 1979). One of the nice outcomes of that project was that we demonstrated a model structure that could be applied successfully based on field-measured parameters (Beven *et al.* 1984). I also learned that parameter optimisation would not necessarily use the model concepts in the correct way (see, for example, Figure 14 in Beven & Kirkby 1979).

My experience with these different types of models proved valuable in being appointed (actually as a 'mathematic modeller' despite my geography degree) at the Institute of Hydrology (IH) in Wallingford in 1977. Part of my time was devoted to the SHE (Système Hydrologique Européen) project, a joint initiative with the Danish Hydraulics Institute (DHI) and SOGREAH in France, funded by a European Community loan. This was another attempt at producing a complete 'physically-based' hydrological model and was led by Mike Abbott who had successfully dealt with the numerical issues of solving the shallow water equations in hydraulics which were the basis of the DHI MIKE series of simulation packages. Before I joined IH, I had participated in the first SHE meeting at Wallingford, and the minutes record that, as a result of my PhD modelling experience, I raised many of the problems that would be met in the SHE project particularly in the decoupling of the saturated and unsaturated zone solutions. This was a pragmatic decision to reduce dimensionality, based on the available computer resource, but was the main reason why it was 1986 before the first SHE applications appeared (Abbott *et al.* 1986a, 1986b; Bathurst 1986). It was later relaxed as computer power increased and much later both the MIKE-SHE and SHETRAN versions of SHE were implemented with fully 3D partially saturated Darcy–Richards subsurface solutions (see Ewen *et al.* 2000; Graham & Butts 2005). Speed could still be an issue, however, and MIKE-SHE has also been used with conceptual groundwater storage components in some applications (see the history of SHE in Refsgaard *et al.* 2010).

After 3 years in Wallingford, in 1979, I moved to the University of Virginia and was able to return to working with Topmodel. I took advantage of its computational speed and the availability of a CDC6600 mainframe computer to start making Monte Carlo runs of the model in around 1980. This soon showed that there were many runs of the model with different parameter sets that gave more or less equivalent results, something that was later developed into the equifinality concept (Beven 1993, 2006a) though equifinality had already been mentioned in my PhD thesis (Beven 1975). It was also the origin of the Generalised Likelihood Uncertainty Estimation (GLUE) methodology, although I did not have the confidence to publish this until much later (Beven & Binley 1992). Returning to Wallingford in 1982, I told the Director, Jim McCulloch, that I would not work on SHE but there was also funding for another physically based model the IH Distributed Model or IHDM that had been started by Liz Morris. We rewrote the IHDM, producing Version 4, that was based on a finite element rather than finite difference methods. It was therefore rather similar to my PhD model but with better numerics and finer discretisations because of more computer resource. The numerics of the IHDM were later improved still further by Calver & Wood (1989) but remained subject to the problems of using the Richards equation as a representation of flows in real soils (see, for example, Beven 1989, a paper that started out as a commentary on the first 1986 SHE applications).

In 1985, I moved to Lancaster and continued work on 3D finite element modelling with Andy Binley (e.g. Binley *et al.* 1989a, 1989b; Binley & Beven 1992); Topmodel and the development of Dynamic Topmodel with Jim Freer (Beven & Freer 2001); modelling flow and transport for water quality (e.g. Page *et al.* 2007; Dean *et al.* 2009; Hollaway *et al.* 2018a); pollutant dispersion and flood forecasting using the Data-Based Mechanistic (DBM) methods of Peter Young (e.g. Wallis *et al.* 1989; Young & Beven 1994); and a wide range of applications of the GLUE methodology (e.g. Beven 2009, 2016a, and the references therein). Some of that work proved controversial, in particular about whether informal likelihood measures and rejection criteria could replace formal statistical methods in model evaluation (see, for example, Beven 2006b, 2008; Andréassian *et al.* 2007; Hall *et al.* 2007; Todini & Mantovan 2007). However, controversy encourages harder thinking about what is important and what is required to go beyond the norms of the current paradigm and make real advances.

This background frames the comments about how to make advances in hydrological modelling that are set out in the following sections. This essentially updates the final chapter of Beven (2012a). I concentrate on what I see as the three most important issues. These are: how to decide whether a model is fit for purpose; how to improve process representations in hydrological models; and how to take advantage of Models of Everywhere. Underlying all those issues, however, is the fundamental problem of improving the hydrological data available for both forcing and evaluating hydrological models.

THE NEED TO IMPROVE HYDROLOGICAL DATA FOR MODEL APPLICATIONS

Hydrological data are highly uncertain (see, most recently, Beven 2019). This is true for the most basic of quantities, such as rainfall at a point, discharge at a point (particularly at the highest and lowest flows), and actual evapotranspiration fluxes (sort of at a point). It is even more problematic if we are interested in the water balance over a catchment area because there are uncertainties in catchment area rainfall, snowfall, evapotranspiration fluxes, and storages. The issue is greater because, in general, the uncertainties involved are the result of a lack of knowledge (i.e. *epistemic* uncertainties) rather than random variability (the *aleatory* uncertainties) (see, for example, Kauffeldt *et al.* 2013; Beven 2016a; Westerberg *et al.* 2016; Wilby *et al.* 2017). In some cases, we choose to treat data uncertainties *as if* they are aleatory because of the convenience of the statistical techniques available (e.g. kriging for the interpolation of areal rainfalls; repeat measurements for ADCP (Accoustic Doppler Current Profiler) estimates of flows; and the choice and fitting of flood frequency distributions). A good example is the use of statistical regression for fitting rating curves for the conversion of observed water levels to discharges. It is often assumed that some simple power law will hold over the range of the data (with or without an offset and with or without multiple segments). This might be satisfactory within the range of the actual gaugings, at least if they are not too variable and if effects such as weed growth and a mobile bed are negligible, but in some cases such extrapolations can be potentially misleading (see, for example, Beven *et al.* 2012; McMillan & Westerberg 2015; Hollaway *et al.* 2018b; and the comparison of Kiang *et al.* 2018).

There is some movement towards the use of extrapolations based on hydraulic modelling of a gauging site, particularly for overbank flows. The review of the Sheepmount rating curve at Carlisle after the 2005 flood is a good example. Consulting engineers were commissioned by the Environment Agency to revisit the rating curve at this site using hydraulic modelling, since the recorded water level was over a metre higher than the highest measured discharge. This led to a significant increase in the estimated discharge relative to that produced by extrapolation of the rating curve fitted to the discharge measurements. The revised rating was then used to estimate the even higher flood peak from Storm Desmond in 2015. However, such estimates are very dependent on the estimation of effective roughness coefficients for the out-of-bank conditions, which is necessarily uncertain. Extrapolated discharge estimates are still often cited without any associated uncertainty range even though there is evidence that effective roughness might be model structure dependent and vary with peak magnitude (e.g. Romanowicz & Beven 2003; Pappenberger *et al.* 2006).

These experiences led to Beven *et al.* (2011) and Beven & Smith (2015), suggesting that some catchment data might be *disinformative* in deciding whether a model is acceptable or not. They identified events that gave exceedingly high or exceedingly low runoff coefficients in a rapid response catchment in the north of England. Clearly if a model is constrained by mass balance, but the data for an event suggest a runoff coefficient greater than 1, then the model is going to produce residuals that reflect the deficiencies in the original data, not only from any failure of the model (see also the examples in global data sets included in Kauffeldt *et al.* 2013). In this case, the problem is quite evident, and if such data are included in model evaluation will lead to bias in inference about parameter values and in predicted outcomes, especially if simple evaluation measures based on the sum of squared errors are used. There will, however, be many other periods of data when the effects on model evaluation will be subtle and difficult to allow for.

The conclusion of this is that we need to be much more careful about considering the value of the available data in model evaluation, and that we need better observational techniques, not only for the inputs and outputs in the water balance equation but also for internal state variables. In the latter case, there is still a great deal of epistemic uncertainty about subsurface flow pathways on hillslopes (and in valley bottoms). Where internal state data are used, there can also be incommensurability between observed variables and simulated variables (e.g. soil moisture at a point relative to the soil moisture output at the discrete element scale of a distributed model). There have been some advances, such as the COSMOS measurement of soil moisture over an area, but that has both variable effective depths and areal extent depending on the levels of near-surface moisture (Zreda *et al.* 2012; Evans *et al.* 2016; Baroni *et al.* 2018).

We also know enough from tracing experiments and the nature of the physics to conclude that the Richards equation should not be used in modelling flow through soils (in fact, we argued this in Beven 1989; Binley *et al.* 1989a, 1989b, nearly 30 years ago). It is based on the wrong experiment that excluded the possibility of preferential flows in focusing on capillary equilibrium conditions. This might be more applicable under relatively dry conditions, but even then, the physics itself suggests that the usual form should not be used if there is any heterogeneity of soil properties within the scale of the application, which is, of course, *always* the case (see also Beven & Germann 1982, 2013; Beven 2012a, 2018b). However, we have no good (non-destructive) measurement techniques at scales of interest with which to study vertical and downslope preferential flows and recharge. Those detailed observations that have been done have suggested that the flows can be highly localised, highly variable, and subject to complex connectivity issues in space and time (e.g. Freer *et al.* 1997; Jencso *et al.* 2009; McGuire 2010; Klaus & Jackson 2018).

In fact, we are not interested in such detail (except in terms of scientific understanding), and it might be better to develop new measurement techniques at larger scales that would integrate over the detail. If, for example, we had an effective and affordable gravity anomaly technique for total water storage over an area; coupled with a method for measuring stream discharges that were sufficiently accurate to determine incremental discharges downstream in a river network, then we might be able to infer much more useful process relationships than those we have currently. However, as a community, we have not been at all pro-active about deciding on priorities for measurement requirements and commissioning new techniques.

The satellite community have done so much more effectively (including the SWOT launch planned for 2021 which will be of some hydrological interest), but, from a hydrological point of view, satellite imaging has always had potential but not actually been that useful, apart from generating digital terrain data, particularly LIDAR (Light Direction and Ranging) that has led to significant improvements in, for example, flood inundation mapping. Even then, however, there are both aleatory and epistemic uncertainties associated with the treatment of the digital numbers (how to deal with vegetation and buildings; small-scale features such as walls and hedges on flood plains; later infilling of sinks or burn-in of channels in the terrain to get consistent flow lines; determination of catchment boundaries etc.) that will have an effect on any model outputs when compared with observations. Most other remote sensing is also associated with epistemic uncertainties, including rainfall and soil moisture estimation, with the result that it provides only some qualitative and uncertain indication of patterns in the landscape relevant to hydrology.

Improving the quantity and quality of hydrological data is essential to what follows, in particular in deciding on whether particular models might be fit-for-purpose. Note that this paper is about how to make improvements in hydrological simulation models. It is not about models used for forecasting, i.e. modelling using data assimilation for getting the best real time n step ahead predictions with minimal uncertainty (see Beven & Young 2013 for a discussion of different types of model prediction). Forecasting does not necessarily require process representations nor physical constraints such as mass balance that may not be a feature of the available data. This is particularly true in forecasting flood events when there may be poor sampling of the most intense rainfalls, and the discharge rating curve may be subject to epistemic uncertainties. Data assimilation is then a valuable tool in improving forecasts. Far better to forecast levels and use data assimilation to compensate for the limitations in the input data (see, for example, Romanowicz *et al.* 2006; Leedal *et al.* 2010). (Note that while I consider data assimilation to be essential in forecasting, I do not consider it to be good practice to use data assimilation to compensate for model deficiencies in simulation modelling, especially if there is no attempt to learn from the data assimilation about how a model might be in error. There have been a number of such studies in the literature. Clearly, it is not possible to use data assimilation to compensate for model deficiencies in simulating the impacts of future changes. It is better than to attempt to produce a realistic estimate of the associated uncertainties, both aleatory and epistemic.)

Here I shall be interested in the representation and simulation of hydrological processes in the context of not only reproducing historical behaviours but also future behaviours under change. Even a cursory survey of the literature will reveal that this is a challenge and difficult to achieve. Hydrological systems are complex and nonlinear, and we have little in the way of techniques for studying patterns of processes at the catchment scale. We rely on the way in which catchments act as integrators over small-scale complexity and heterogeneity in resorting to the calibration of simple model representations against the very discharge data that we want to predict. That clearly helps in getting better reproduction of discharges without change but not necessarily for the right reasons. Getting good results for the wrong reasons could then be misleading when we want to simulate the impacts of change (rarely is any consideration given to change during a calibration period, but see Merz *et al.* 2011; Peel & Blöschl 2011; Harrigan *et al.* 2014). In the past, I have had some success in making predictions using only measured parameters (e.g. Beven *et al.* 1984), but also some notable failures (e.g. Beven 2001).

EVALUATING HYDROLOGICAL MODELS AS FIT-FOR-PURPOSE

We know very well that the process representations used in hydrological models are only approximations to the real-world complexity of surface and subsurface flows. It is also obvious that the epistemic issues with hydrological data mean that we would not expect even a perfect model to provide perfect predictions. We see this in the comparisons of observed and predicted variables in a multitude of academic papers and reports to clients. Sometimes, indeed, it seems that the predictions are rather poor, especially if models are applied without calibration as if to an ungauged catchment. Calibration is generally helpful in finding parameter sets that give predictions that are closer to the observations at least in the calibration period. When a split record evaluation is also done, it is common to find that the model performance is not so good in the validation period or under different seasonal or climate conditions (Refsgaard & Knudsen 1996; Freer *et al.* 2003; Choi & Beven 2007; Coron *et al.* 2014; Fowler *et al.* 2016, 2018; Dakhlaoui *et al.* 2017; Pool *et al.* 2017). This might be the result of over-fitting an overparameterised model; it might be because the model is producing good results in calibration for the wrong reasons; it might be only because the forcing data errors are quite different in the validation period. For more severe testing (see Klemes 1986; Refsgaard & Knudsen 1996; Ewen & Parkin 1996; Seibert 2003), it is often difficult to declare any form of success.

We can think about models as hypotheses about how a hydrological system functions (e.g. Beven 2012b, 2018a). Thus, testing whether a model should be considered as fit-for-purpose can be considered a form of hypothesis testing, with the possibility of rejecting models that do not fit the evaluation data to some defined level of acceptability. Model rejection in this sense is a good thing; it means that we need to make some improvements, either to the model structure or to the data that we are using with the model (Beven 2018a). Clearly, methods for hypothesis testing are well developed in statistics under assumptions that variables can be considered to have aleatory variability. However, when we *know* that we are dealing with epistemic uncertainties, it might be incoherent to use simple statistical assumptions (e.g. Beven *et al.* 2008). This is evident, for example, in the use of formal likelihood functions in model evaluation that, particularly for long time series, can give quite a misleading impression of the relative merits of different models and parameter sets (e.g. Beven & Smith 2015; Beven 2016a).

In assessing fitness-for-purpose, of course, we do need to consider what is the purpose. We can differentiate between two major types of purpose (though each could have a variety of subdivisions). The first is in the use of models to test the science, i.e. the understanding of how a hydrological system might function. This might involve the more detailed consideration of the internal states and other detail in experimental plots and catchments, and how they differ from responses reported from elsewhere. The second is in the use of models for decision-making. The important factor then is that the model should make predictions of the future behaviour of a hydrological system that will not deviate too far from what would happen under the assumed boundary conditions. This might allow a greater degree of approximation to be considered to be acceptable, especially if decisions are being taken at larger scales (such as in the methods used for the UK National Flood Risk Assessment that is currently under revision). A particular feature of this second purpose is that the results cannot really be tested, even if a model has survived a validation test, since the future boundary conditions are necessarily unknown or epistemically uncertain (see, for example, the post-audit analyses of groundwater models in Konikow & Bredehoeft (1992), where some models failed only because of poor assumptions about the future boundary conditions). We might hope, of course, that as the science evolves, the purpose of improving understanding will feed into the purpose of decision-making, with a better theoretical basis for moving from local scales to national scales and for assessing changes in parameter values but we are not there yet (see below).

The question remains of how should we test models as hypotheses in the face of epistemic uncertainties? Beven & Lane (2019) suggest that one way of looking at this problem is in the form of testing for model invalidation (see also Beven 2018a). There is, of course, a long history of applying such tests, at least implicitly in the form of not invalidating a model based on its simulated outputs. Every time a referee accepts a paper with model results for publication, s/he is essentially applying such a test. Every time a report is presented to a client, then the authors of that report have applied such a test. Every time a report is accepted by the client (perhaps after an independent assessment by another consultant), then such a test has been applied. Most of these judgments are qualitative and subjective, albeit that they may be supported by some quantitative measures (such as quoting the Nash–Sutcliffe efficiency despite all of its faults as a measure of calibration or validation performance).

It is therefore interesting to speculate about what information such a group of experts would require in order to make such an invalidation more rigorous, both in the use of models for predicting an ungauged catchment and in the case where some output observations are available to evaluate model runs. One interesting feature of this strategy is that there is a possibility for the users of the model outputs, such as decision or policymakers or stakeholders affected by a decision, to be involved in such a process in considering not only the acceptability of the model outputs but also the assumptions that contribute to the outputs (see Beven 2018a, and the condition tree approach of Beven *et al.* 2014).

There is actually a precedent for this type of approach in the 'blind validation' approach of Ewen & Parkin (1996). This requires the modeller (in their case) to define some criteria for acceptability prior to making any model runs. Model parameters were estimated from the past experience, and no prior model calibration was allowed. The range of simulated outcomes was then compared with available observations of flows and internal state data (assumed at that time to be known accurately). Blind validation was applied to the SHE model by Parkin *et al.* (1996) and Bathurst *et al.* (2004). In both cases, the model simulations failed to meet all the defined validation criteria. In the application of Parkin *et al.* (1996), the model failed one out of four tests; in the case of Bathurst *et al.* (2004), two out of 10 tests were failed. This was despite the criteria for success being rather relaxed and some model simulations being excluded on the basis of expert evaluations. These failures do not seem to have had much effect on the use of the SHE model elsewhere. In fact, the failures are not mentioned at all in the SHE review paper of Refsgaard *et al.* (2010), which includes just a brief passing mention of the development of model testing methods based on the Klemes (1986) concepts. There have been no other applications of this blind validation methodology, to my knowledge, though it has much in common with the setting of limits of acceptability within the GLUE methodology (see Beven 2006a, 2009, 2016a) that has led to some other model invalidations (e.g. Page *et al.* 2007; Dean *et al.* 2009; Liu *et al.* 2009; Hollaway *et al.* 2018a).

One of the issues in this type of evaluation is, again, the data being used to both drive and test a model as hypothesis. Since we do not expect a model to be better than the data it is used with, any invalidation test should first make some assessment and allowance for the uncertainties, both epistemic and aleatory, associated with those data, although in some (wet) cases any model that gets the water balance separation approximately correct might provide quite good measures of performance (e.g. Seibert *et al.* 2018). How uncertain do we expect the inputs used to force the model to be? If we have observations of the system response, how uncertain are those observations relative to the variables predicted by the model? We do not expect this assessment of uncertainty to be a simple statistical variability (though lacking better knowledge we might choose to treat it as such). We are not used to framing model testing in this way (and indeed perhaps, we have avoided it because these are very difficult questions to resolve when we expect the nature of errors in the inputs to vary from event to event, and parameter interactions to be complex). Data uncertainty also raises the issue of how to avoid Type I hypothesis testing errors (accepting a model hypothesis that is not fit-for-purpose because of the data uncertainties) and Type II errors (rejecting a model hypothesis that would be fit-for-purpose because of the data uncertainties). The former is more problematic but should hopefully be reduced as new data, or different types of data are added to the assessment. Such difficulties should not, however, stop us from thinking more deeply about how to make an invalidation test more rigorous.

A further feature that might be considered is whether a model contradicts some secure evidence on the nature of the system response. If that is the case, it should not be considered as fit-for-purpose. We want to base decisions on predictions from a model that, as far as possible, is producing the right results for the right reasons. A nice example of this appears in the very first Topmodel paper (Beven & Kirkby 1979) where it was shown that optimising the model parameters resulted in using the model structure in a way that contradicts the theory on which it was based by using the subsurface store with a very low time constant to control the timing of fast runoff. There are also examples from other domains such as climate models (e.g. Liepert & Lo 2013). Thus, how to show that a model is giving the right results for the right reasons should be a subject for some deeper thought (see, for example, Kirchner 2006).

An interesting possibility that arises from applying more rigorous testing to model applications in hydrology is that all the models tried might be rejected as fit-for-purpose. This invalidation might be for different parameter sets in a single model structure; it might extend to multiple model structures. There are published examples of where all the models tried have been rejected (see, most recently, the case of the SWAT model in Hollaway *et al.* (2018a), in an application to a small UK catchment). As noted earlier, such model rejection is really a good outcome, in that it requires either that we do better modelling or find better data, or that we find some other way of making decisions within an adaptive management framework. We should, note, however, that even where more rigorous invalidation testing is carried out, the results will always be conditional on the information that is to hand now. The future remains epistemically uncertain, and the possibility of future surprise remains. That should not, however, be a reason for relaxing the testing. It should still be considered as a poor practice to relax rejection criteria just because a decision needs to be made. That may not result in a good decision if the model is not fit-for-purpose or if the decision is sensitive to the uncertainty in model predictions.

IMPROVING PROCESS REPRESENTATIONS IN HYDROLOGICAL MODELS

The concept of being able to reject models as hypotheses has an important implication; that we might be able to learn from the nature of the rejection to refine the representation of hydrological processes and systems where this is shown to be necessary. In this context, model rejection is a good outcome. It is the starting point for where the creativity of analysis and thought is required for doing better in the future.

It is already possible, however, to make some suggestions as to what such innovations might look like, particularly if we want process representations that will satisfy the needs to predict both flow and transport within a consistent framework. This assumes greater importance when we start to accept the limitations of gradient-based continuum approaches such as the Buckingham–Richards equation (which I have argued need to be reconsidered since Beven 1989). Such a framework is required to consider both velocities (in predicting conservative transport) and celerities (in predicting flows). Since celerities are generally different and faster than velocities, it follows that any process representation should be length scale-dependent, i.e. different scales of spatial discretisation might require different parameter values. The difference between velocities and celerities will also be state-dependent, suggesting that, at any scale, the hysteresis on the storage–flux response will change with system state. This has been shown numerically using the Multiple Interacting Pathway (MIP) model by Davies & Beven (2015).

The MIPs model allows velocity distributions to be specified as part of a random particle representation of all the water in the flow domain. Celerities follow from the filling and emptying of storage in the system. It is a computationally expensive modelling strategy and therefore has, to date, been restricted to small-scale applications. While there is still much to explore in the interaction between the scale of discretisation, time step, velocity distributions, and transition probabilities, it does have the type of consistent framework that might be valuable in future. Zehe & Jackisch (2016) and Jackisch & Zehe (2018) have taken a somewhat similar approach including more explicit consideration of the effects of capillarity. Such approaches might be one way of approaching a theory of scale-dependent process representations for both flow and transport.

I have argued before (e.g. Beven 2006c, 2012a) that there is already a useful framework within which new process representations might be embedded. This is the Representative Elementary Watershed (REW) framework (see, for example, Reggiani *et al.* 2000; Reggiani & Schellekens 2003). This sets out a framework of mass, energy, and momentum equations that is common for any spatial discretisation. However, those balance equations need closure, i.e. a way of defining the flux terms of mass, energy, and momentum at the boundaries of each discrete element, together with how those fluxes depend on the internal states of the system. I believe that this will lead to closure schemes based on hysteretic relationships between element storages and boundary fluxes. A move in this direction would, of course, be greatly enhanced by the availability of the relevant storages or fluxes at the element scale, and it may be (again) that real progress will await the availability of new measurement techniques. What we should not do, however, is to continue to ignore the implications of the difference between velocities and celerities and the scale-dependent and hysteretic nature of hydrological responses at the element scale.

It is perhaps worth pointing out that the asymmetry of the unit hydrograph or linear transfer functions derived at catchment scales is a representation of hysteresis in the storage–flow relationship. But as a linear model, it relies on a way of processing the inputs to represent the effects of nonlinearity and antecedent conditions in predicting the catchment response at a wider range of conditions. I could speculate that if input, storage and output data were available for discrete elements of the landscape (or arbitrary REWs), then a transfer function modelling framework such as the DBM approach developed by Young (1998) and Young & Beven (1994) would be a suitable way of deriving closure schemes at the required scale. The parameters of such a model would then be quite different from those we use today: the time constants for the linear transfer function and some coefficients for nonlinear processing the input sequence. Given additional tracer data, it might also be possible to derive a consistent set of concepts relating parameters for both flow and transport within such a framework (e.g. Harman 2019). The emphasis is, again, on making the right type of data available, initially at research locations, so that we can learn about how to produce closure schemes that might be applicable more widely.

However, I could also speculate that rather than accepting a limitation to the linear transfer function or unit hydrograph, with its constant time distribution of contributions of effective rainfall to the hydrograph, perhaps there will be other ways of analysing such data that might more explicitly reflect antecedent states and input intensities at the required scale of discretisation. There are methods of developing hysteretic functions that have been applied to hydrological systems (e.g. O'Kane & Flynn 2007; Appelbe *et al.* 2009), but these also have some rather strong assumptions. Given recent developments in data-mining techniques, might this be a way of deriving the *forms* of functions that would be applicable more widely, that would suggest quite different process representations than those being used today?

HYDROLOGICAL MODELS OF EVERYWHERE

The other advance that is certainly going to have a major impact on modelling practice is the much more widespread availability of spatial predictions of hydrological models on the internet. I first suggested a Models of Everywhere concept more than a decade ago (Beven 2007; Beven & Alcock 2012), but it is only relatively recently that this has become computationally easier to implement and computer scientists have become more interested in the problem of producing facilitating software (e.g. Blair *et al.* 2019).

What is critical to this Models of Everywhere concept is that the predictions are a sufficiently fine resolution that local stakeholders can relate to them directly. The concept is therefore quite different from providing the global 'hyperresolution' simulations presented, for example, by Wood *et al.* (2011). Hyperresolution in their sense is of the order of 1 km (see Bierkens *et al.* 2015) and while there may be some variables that local stakeholders can relate to at that scale, there will also be a great deal of hyperresolution ignorance about what parameters and variables at that scale might mean (see, for example, the discussion in Beven *et al.* 2015). There is a movement to finer-resolution, continental-scale simulations, such as the HydroBlocks of Chaney *et al.* (2016), which is based on the Dynamic Topmodel. At much finer scales, such as the 2 m scale used in producing the UK pluvial flooding maps, the ability of people with local knowledge to provide feedback on the model outputs is much more direct. In this case, modelling becomes much more of a learning process, driven by the feedback about where the model predictions are demonstrably wrong. It is a learning process about *places* that starts to reflect the uniqueness of places in terms of both learning about appropriate effective parameter values and learning about appropriate process representations (see Beven 2000). The possibility of local feedback on the acceptability of model simulations will change the nature of the modelling process in fundamental ways. While we might start with general model structures that are applied to places as in the past, what we need are methods of learning about places from the availability of local data, effective ways of obtaining new data for different purposes, and making use of local (perhaps qualitative) knowledge and expertise (see, for example, the study of Landström *et al.* 2011).

There is an interesting issue in the question of data assimilation in applications of Models of Everywhere. Clearly, we would wish to use all the useful information available to test models locally and to ensure that we get the right results for the right reasons. This might include whatever quantitative data might be available but might also be a matter of learning how to use 'soft' data in model evaluations (see, for example, Seibert & McDonnell 2002; Fenicia *et al.* 2008; Winsemius *et al.* 2009). We would also like to re-evaluate models as more data are made available. But, as noted earlier, there have been cases where data assimilation is used simply to compensate for model deficiencies by updating model states so as to get a better predicted outcome. If the purpose of modelling is real-time forecasting into the near future, then that might be acceptable or even advisable. Where the purpose is for simulation and assessing the impacts of future change, then we should be very wary of compensating for important model deficiencies. For Models of Everywhere, we might want to do both forecasting and simulation, in which case it will be important to learn from the process of data assimilation for forecasting in improving the model formulation for simulation. There have been few studies (to my knowledge) that have done so (but see the learning from nonstationarity of Westra *et al.* 2014, as an example of the type of analysis that might lead to model modifications). More generally, forecasters have been satisfied with using data assimilation to get better forecasts, and simulation modellers have been satisfied with using calibration to either find an optimal model or constrain the associated uncertainty. Perhaps, we can do better or at least be a little more thoughtful in applying models. The feedback from users once Models of Everywhere visualisations are more widely available may force us to do so.

CONCLUSIONS

I have written about these issues in many past papers (including Beven 2016b), but this has been a useful opportunity to bring the strands of thought about the future of hydrological modelling in one place. I do think that hydrology remains a field of inexact science that is still greatly constrained by observational limitations, and it would be really good to see the community make a real effort to decide on what its priorities should be and then move to commission what is needed (as has happened, for example, with the SWOT satellite). The process might be long, but the benefits to the science would be great, including for testing models as hypotheses, developing new process representations, and constraining predictive uncertainties.

The role of Models of Everywhere in improving modelling capability will also make for an interesting future. What new techniques for learning about places and for learning from clear errors in representing the response of places will need to be developed? And how can new types of knowledge be used to constrain uncertainties? What should the learning framework for both quantitative and qualitative information look like, including the issue of distinguishing information from disinformation. These are issues that are relevant to a wider range of research areas than hydrology which is just one of many inexact environmental sciences (Beven 2002, 2019).

There is a particularly interesting aspect of uncertainty for the modeller in this context. A realistic assessment of uncertainty in predicting how places respond will mean that the modeller is much less likely to be obviously wrong in those predictions. This is clearly a good thing (at least from a modeller's point of view), but should not preclude an effort being made to carry out model testing and find ways of reducing that predictive uncertainty.

As I said in the talk on which this paper is based, I am ending my career with much more uncertainty than when I started as a young PhD student in 1971. But that is a good thing – it means that there is still so much good research to do in the closely linked areas of novel observational methods, closure schemes and model testing, theoretical development, and learning about places. In particular, learning about the assessment of epistemic uncertainties will also lead to the development of methods for reducing those uncertainties. The near future could be an exciting time for hydrological research and practice.

ACKNOWLEDGEMENTS

Over a long career, I have been fortunate to work and collaborate with many excellent hydrological modellers and experimentalists; a number that is really too long to list here. I will just mention the contribution of Dr Peter Metcalfe who had only recently started work on the Q-NFM project led by Dr Nick Chappell at Lancaster University before his sudden death in a climbing accident. Working with Peter on the problem of modelling distributed natural flood management measures in this project (NERC grant no. NE/R004722/1) was instrumental in my thinking again about how to improve hydrological models. I am also grateful to Jan Seibert and two anonymous referees who made some useful suggestions for relevant papers and improvements to the paper and presentation.

REFERENCES

Abbott, M. B., Bathurst, J. C., Cunge, J. A., O'Connell, P. E. & Rasmussen, J. 1986a An introduction to the European Hydrological System-Systeme Hydrologique Europeen, 'SHE', 1: history and philosophy of a physically-based, distributed modelling system. *Journal of Hydrology* **87** (1–2), 45–59.

Abbott, M. B., Bathurst, J. C., Cunge, J. A., O'connell, P. E. & Rasmussen, J. 1986b An introduction to the European Hydrological System-Systeme Hydrologique Europeen, 'SHE', 2: structure of a physically-based, distributed modelling system. *Journal of Hydrology* **87** (1–2), 61–77.

Andréassian, V., Lerat, J., Loumagne, C., Mathevet, T., Michel, C., Oudin, L. & Perrin, C. 2007 What is really undermining hydrologic science today? *Hydrological Processes* **21** (20), 2819–2822.

Appelbe, B., Flynn, D., Mcnamara, H., Philip, O. K., Pimenov, A., Pokrovskii, A., Rachinskii, D. & Zhezherun, A. 2009 Rate-independent hysteresis in terrestrial hydrology. *IEEE Control Systems* **29** (1), 44–69.

Baroni, G., Scheiffele, L. M., Schrön, M., Ingwersen, J. & Oswald, S. E. 2018 Uncertainty, sensitivity and improvements in soil moisture estimation with cosmic-ray neutron sensing. *Journal of Hydrology* **564**, 873–887.

Bathurst, J. C. 1986 Physically-based distributed modelling of an upland catchment using the Systeme Hydrologique Europeen. *Journal of Hydrology* **87** (1–2), 79–102.

Bathurst, J. C., Ewen, J., Parkin, G., O'Connell, P. E. & Cooper, J. D. 2004 Validation of catchment models for predicting land-use and climate change impacts. 3. Blind validation for internal and outlet responses. *Journal of Hydrology* **287** (1–4), 74–94.

Beven, K. J. 1975 *A Deterministic Spatially Distributed Model of Catchment Hydrology*. PhD Thesis, University of East Anglia, Norwich.

Beven, K. J. 1989 Changing ideas in hydrology: the case of physically based models. *Journal of Hydrology* **105**, 157–172.

Beven, K. J. 1993 Prophecy, reality and uncertainty in distributed hydrological modelling. *Advances in Water Resources* **16**, 41–51.

Beven, K. J. 2000 Uniqueness of place and process representations in hydrological modelling. *Hydrology and Earth System Sciences* **4** (2), 203–213.

Beven, K. J. 2001 Dalton medal lecture: how far can we go in distributed hydrological modelling? *Hydrology and Earth System Sciences* **5** (1), 1–12.

Beven, K. J. 2002 Towards a coherent philosophy for environmental modelling. *Proceedings Of The Royal Society Of London A* **458**, 2465–2484.

Beven, K. J. 2006a A manifesto for the equifinality thesis. *Journal of Hydrology* **320**, 18–36.

Beven, K. J. 2006b On undermining the science? *Hydrological Processes* **20**, 3141–3146.

Beven, K. J. 2006c The Holy Grail of Scientific Hydrology: as closure. *Hydrology and Earth Systems Science* **10**, 609–618.

Beven, K. J. 2007 Working towards integrated environmental models of everywhere: uncertainty, data, and modelling as a learning process. *Hydrology and Earth System Science* **11** (1), 460–467.

Beven, K. J. 2008 On doing better hydrological science. *Hydrological Processes* **22**, 3549–3553.

Beven, K. J. 2009 *Environmental Modelling - An Uncertain Future*. Routledge, London.

Beven, K. J. 2012a *Rainfall-Runoff Modelling - The Primer*, 2nd edn. Wiley-Blackwell, Chichester.

Beven, K. J. 2012b Causal models as multiple working hypotheses about environmental processes. *Comptes Rendus Geoscience, Académie des Sciences* **344**, 77–88. doi:10.1016/j.crte.2012.01.005.

Beven, K. J. 2016a EGU Leonardo Lecture: Facets of Hydrology – epistemic error, non-stationarity, likelihood, hypothesis testing, and communication. *Hydrological Sciences Journal* **61** (9), 1652–1665. doi:10.1080/02626667.2015.1031761.

Beven, K. J. 2016b Advice to a young hydrologist. *Hydrological Processes* **30**, 3578–3582. doi:10.1002/hyp.10879.

Beven, K. J. 2018a On hypothesis testing in hydrology: why falsification of models is still a really good idea. *WIRES Water*. doi:10.1002/wat2.1278.

Beven, K. J. 2018b A century of denial: preferential and non-equilibrium water flow in soils, 1864–1984. *Vadoze Zone Journal* **17**, 180153, https://dx.doi.org/10.2136/vzj2018.08.0153.

Beven, K. J. 2019 Towards a methodology for testing models as hypotheses in the inexact sciences. *Proceedings Royal Society A* **475** (2224), doi: 10.1098/rspa.2018.0862.

Beven, K. J. & Alcock, R. 2012 Modelling everything everywhere: a new approach to decision making for water management under uncertainty. *Freshwater Biology* **56**, 124–132. doi:10.1111/j.1365-2427.2011.02592.x.

Beven, K. J. & Binley, A. M. 1992 The future of distributed models: model calibration and uncertainty prediction. *Hydrological Processes* **6**, 279–298.

Beven, K. J. & Callen, J. L. 1979 'HYDRODAT: a system of FORTRAN computer programs for the preparation and analysis of hydrological data from charts. British Geomorphological Research Group. *Technical Bulletin* **23**, Norwich, UK.

Beven, K. J. & Freer, J. 2001 A dynamic TOPMODEL. *Hydrological Processes* **15** (10), 1993–2011.

Beven, K. J. & Germann, P. 1982 Macropores and water flow in soils. *Water Resources Research* **18** (5), 1311–1325.

Beven, K. J. & Germann, P. F. 2013 Macropores and water flow in soils revisited. *Water Resources Research* **49** (6), 3071–3092. doi:10.1002/wrcr.20156.

Beven, K. J. & Kirkby, M. J. 1979 A physically-based variable contributing area model of basin hydrology. *Hydrological Sciences Bulletin* **24** (1), 43–69.

Beven, K. J. & Lane, S. 2019 Invalidation of models and fitness-for-purpose: a rejectionist approach. In: *Computer Simulation Validation – Fundamental Concepts, Methodological Frameworks, and Philosophical Perspectives* (C. Beisbart & N. J. Saam, eds.). Springer, Cham, pp. 145–171.

Beven, K. J. & Smith, P. J. 2015 Concepts of information content and likelihood in parameter calibration for hydrological simulation models. *ASCE Journal of Hydrologic Engineering*. doi:10.1061/(ASCE)HE.1943-5584.0000991.

Beven, K. & Young, P. 2013 A guide to good practice in modeling semantics for authors and referees. *Water Resources Research* **49** (8), 5092–5098. doi:10.1002/wrcr.20393.

Beven, K. J., Kirkby, M. J., Schofield, N. & Tagg, A. 1984 Testing a physically-based flood forecasting model (TOPMODEL) for three UK catchments. *Journal of Hydrology* **69**, 119–143.

Beven, K. J., Smith, P. J. & Freer, J. 2008 So just why would a modeller choose to be incoherent? *Journal of Hydrology* **354**, 15–32.

Beven, K., Smith, P. J. & Wood, A. 2011 On the colour and spin of epistemic error (and what we might do about it). *Hydrology and Earth System Sciences* **15**, 3123–3133. doi:10.5194/hess-15-3123-2011.

Beven, K. J., Buytaert, W. & Smith, L. A. 2012 On virtual observatories and modeled realities (or why discharge must be treated as a virtual variable). *Hydrological Processes*. doi:10.1002/hyp.9261.

Beven, K. J., Leedal, D. T. & McCarthy, S. 2014 *Framework for Assessing Uncertainty in Fluvial Flood Risk Mapping, CIRIA Report C721*. Available from: http://www.ciria.org/Resources/Free_publications/fluvial_flood_risk_mapping.aspx.

Beven, K. J., Cloke, H., Pappenberger, F., Lamb, R. & Hunter, N. 2015 Hyperresolution information and hyperresolution ignorance in modelling the hydrology of the land surface. *Science China Earth Sciences* **58** (1), 25–35.

Bierkens, M. F., Bell, V. A., Burek, P., Chaney, N., Condon, L. E., David, C. H., de Roo, A., Döll, P., Drost, N., Famiglietti, J. S. & Flörke, M. 2015 Hyper-resolution global hydrological modelling: what is next? 'Everywhere and locally relevant'. *Hydrological Processes* **29** (2), 310–320. https://doi.org/10.1002/hyp.10391.

Binley, A. M. & Beven, K. J. 1992 Three-dimensional modelling of hillslope hydrology. *Hydrological Processes* **6**, 347–359.

Binley, A. M., Beven, K. J. & Elgy, J. 1989b A physically-based model of heterogeneous hillslopes. II. Effective hydraulic conductivities. *Water Resources Research* **25** (6), 1227–1233.

Binley, A. M., Elgy, J. & Beven, K. J. 1989a A physically-based model of heterogeneous hillslopes. I. Runoff production. *Water Resources Research* **25** (6), 1219–1226.

Blair, G. S., Beven, K., Lamb, R., Bassett, R., Cauwenberghs, K., Dean, G., Hunter, N., Edwards, E., Nundloll, V., Samreen, F., Simm, W. & Towe, R. 2019 Models of everywhere revisited: A technological perspective. *Environmental Modelling and Software*. https://doi.org/10.1016/j.envsoft.2019.104521

Calver, A. & Wood, W. L. 1989 On the discretization and cost-effectiveness of a finite element solution for hillslope subsurface flow. *Journal of Hydrology* **110** (1–2), 165–179.

Chaney, N. W., Metcalfe, P. & Wood, E. F. 2016 Hydroblocks: a field-scale resolving land surface model for application over continental extents. *Hydrological Processes* **30** (20), 3543–3559.

Choi, H. T. & Beven, K. J. 2007 Multi-period and multi-criteria model conditioning to reduce prediction uncertainty in an application of TOPMODEL within the GLUE framework. *Journal of Hydrology* **332** (3–4), 316–336. doi:10.1016/j.jhydrol.2006.07.012.

Coron, L., Andréassian, V., Perrin, C., Bourqui, M. & Hendrickx, F. 2014 On the lack of robustness of hydrologic models regarding water balance simulation: a diagnostic approach applied to three models of increasing complexity on 20 mountainous catchments. *Hydrology and Earth System Sciences* **18** (2), 727–746. doi:10.5194/hess-18-727-2014.

Crawford, N. H. & Linsley, R. K. 1966 *Digital Simulation in Hydrology: Stanfrod Watershed Model IV, Technical Report 39*. Department of Civil Engineering, Stanford University, CA.

Dakhlaoui, H., Ruelland, D., Tramblay, Y. & Bargaoui, Z. 2017 Evaluating robustness of conceptual rainfall-runoff models under climate variability in Northern Tunisia. *Journal of Hydrology* **550**, 201–217.

Davies, J. & Beven, K. J. 2015 Hysteresis and scale in catchment storage, flow, and transport. *Hydrological Processes* **29** (16), 3604–3615. doi:10.1002/hyp.10511.

Dean, S., Freer, J. E., Beven, K. J., Wade, A. J. & Butterfield, D. 2009 Uncertainty assessment of a process-based integrated catchment model of phosphorus (INCA-P). *Stochastic Environmental Research and Risk Assessment* **23**, 991–1010. doi:10.1007/s00477-008-0273-z.

Evans, J. G., Ward, H. C., Blake, J. R., Hewitt, E. J., Morrison, R., Fry, M., Ball, L. A., Doughty, L. C., Libre, J. W., Hitt, O. E., Rylett, D., Ellis, R. J., Warwick, A. C., Brooks, M., Parkes, M. A., Wright, G. M. H., Singer, A. C., Boorman, D. B. & Jenkins, A. 2016 Soil water content in southern England derived from a cosmic-ray soil moisture observing system – COSMOS-UK: soil water content in southern England – COSMOS-UK. *Hydrological Processes*. doi:10.1002/hyp.10929.

Ewen, J. & Parkin, G. 1996 Validation of catchment models for predicting land-use and climate change impacts. 1. Method. *Journal of Hydrology* **175** (1–4), 583–594.

Ewen, J., Parkin, G. & O'Connell, P. E. 2000 SHETRAN: distributed river basin flow and transport modeling system. *ASCE Journal of Hydrologic Engineering* **5** (3), 250–258.

Fenicia, F., McDonnell, J. J. & Savenije, H. H. G. 2008 Learning from model improvement: on the contribution of complementary data to process understanding. *Water Resources Research* **44**, W06419.

Fowler, K. J., Peel, M. C., Western, A. W., Zhang, L. & Peterson, T. J. 2016 Simulating runoff under changing climatic conditions: revisiting an apparent deficiency of conceptual rainfall-runoff models. *Water Resources Research* **52** (3), 1820–1846. https://doi.org/10.1002/2015WR018068.

Fowler, K., Coxon, G., Freer, J., Peel, M., Wagener, T., Western, A., Woods, R. & Zhang, L. 2018 Simulating runoff under changing climatic conditions: a framework for model improvement. *Water Resources Research*. doi:10.1029/2018WR023989.

Freer, J., McDonnell, J., Beven, K. J., Brammer, D., Burns, D., Hooper, R. P. & Kendal, C. 1997 Topographic controls on subsurface stormflow at the hillslope scale for two hydrologically distinct small catchments. *Hydrological Processes* **11** (9), 1347–1352.

Freer, J., Beven, K. & Peters, N. 2003 Multivariate seasonal period model rejection within the generalised likelihood uncertainty estimation procedure. In: *Calibration of Watershed Models* (Q. Duan, H. Gupta, S. Sorooshian, A. Rousseau & R. Turcotte, eds.), pp. 69–87. doi:10.1029/WS006p0069.

Freeze, R. A. & Harlan, R. L. 1969 Blueprint for a physically-based, digitally-simulated, hydrologic response model. *Journal of Hydrology* **9**, 237–258.

Graham, D. N. & Butts, M. B. 2005 Flexible integrated watershed modelling with MIKE SHE. In: *Watershed Models* (V. P. Singh & D. K. Frevert, eds.). CRC Press, Boca Raton, FL, pp. 245–272.

Hall, J., O'Connell, E. & Ewen, J. 2007 On not undermining the science: coherence, validation and expertise. Discussion of invited commentary by Keith Beven. *Hydrological Processes: An International Journal* **21** (7), 985–988.

Harman, C. 2019 Age-ranked storage-discharge relations – a unified description of spatially-lumped flow and water age in hydrologic systems. *Water Resources Research* **55** (8), 7143–7165.

Harrigan, S., Murphy, C., Hall, J., Wilby, R. L. & Sweeney, J. 2014 Attribution of detected changes in streamflow using multiple working hypotheses. *Hydrology and Earth System Sciences* **18**, 1935–1952.

Hollaway, M. J., Beven, K. J., Benskin, C. M. W. H., Collins, A. L., Evans, R., Falloon, P. D., Forber, K. J., Hiscock, K. M., Kahana, R., Macleod, C. J. A., Ockenden, M. C., Villamizar, M. L., Wearing, C., Withers, P. J. A., Zhou, J. G. & Haygarth, P. M. 2018a Evaluating a processed based water quality model on a UK headwater catchment: what can we learn from a 'limits of acceptability' uncertainty framework? *Journal of Hydrology*. doi:10.1016/j.jhydrol.2018.01.063.

Hollaway, M. J., Beven, K. J., Benskin, C. M. W. H., Collins, A. L., Evans, R., Falloon, P. D., Forber, K. J., Hiscock, K. M., Kahana, R., Macleod, C. J. A., Ockenden, M. C., Villamizar, M. L., Wearing, C., Withers, P. J. A., Zhou, J. G., Barber, N. J. & Haygarth, P. M. 2018b A method for uncertainty constraint of catchment discharge and phosphorus load estimates. *Hydrological Processes* **32**, 2779–2787. doi:10.1002/hyp.13217.

Jackisch, C. & Zehe, E. 2018 Ecohydrological particle model based on representative domains. *Hydrology and Earth System Sciences* **22** (7), 3639–3662.

Jencso, K. G., McGlynn, B. L., Gooseff, M. N., Wondzell, S. M., Bencala, K. E. & Marshall, L. A. 2009 Hydrologic connectivity between landscapes and streams: transferring reach-and plot-scale understanding to the catchment scale. *Water Resources Research* **45**, W04428. doi.org/10.1029/2008WR007225.

Kauffeldt, A., Halldin, S., Rodhe, A., Xu, C.-Y. & Westerberg, I. K. 2013 Disinformative data in large-scale hydrological modelling. *Hydrology and Earth System Sciences* **17**, 2845–2857. https://doi.org/10.5194/hess-17-2845-2013

Kiang, J. E., Gazoorian, C., McMillan, H., Coxon, G., Le Coz, J., Westerberg, I. K., Belleville, A., Sevrez, D., Sikorska, A. E., Petersen-Øverleir, A. & Reitan, T. 2018 A comparison of methods for streamflow uncertainty estimation. *Water Resources Research*. doi.org/10.1029/2018WR022708

Kirchner, J. W. 2006 Getting the right answers for the right reasons: linking measurements, analyses, and models to advance the science of hydrology. *Water Resources Research* **42**, 3. doi:10.1029/2005WR004362.

Klaus, J. & Jackson, C. R. 2018 Interflow is not binary: a continuous shallow perched layer does not imply continuous connectivity. *Water Resources Research*. doi.org/10.1029/2018WR022920

Klemes, V. 1986 Operational testing of hydrological simulation models. *Hydrological Sciences Journal* **31** (1), 13–24.

Konikow, L. F. & Bredehoeft, J. D. 1992 Ground-water models cannot be validated. *Advances in Water Resources* **15** (1), 75–83.

Landström, C., Whatmore, S. J., Lane, S. N., Odoni, N., Ward, N. & Bradley, S. 2011 Coproducing flood risk knowledge: redistributing expertise in critical 'participatory modelling'. *Environment and Planning A* **43** (7), 1617–1633.

Leedal, D. T., Neal, J., Beven, K., Young, P. & Bates, P. 2010 Visualization approaches for communicating real-time flood forecasting level and inundation information. *Journal of Flood Risk Management* **3**, 140–150.

Liepert, B. G. & Lo, F. 2013 CMIP5 update of 'Inter-model variability and biases of the global water cycle in CMIP3 coupled climate models. *Environmental Research Letters* **8**, 029401.

Liu, Y., Freer, J., Beven, K. & Matgen, P. 2009 Towards a limits of acceptability approach to the calibration of hydrological models: extending observation error. *Journal of Hydrology* **367** (1–2), 93–103.

McGuire, K. J. & McDonnell, J. J. 2010 Hydrological connectivity of hillslopes and streams: characteristic time scales and nonlinearities. *Water Resources Research* **46**, W10543. doi:10.1029/2010WR009341.

McMillan, H. K. & Westerberg, I. K. 2015 Rating curve estimation under epistemic uncertainty. *Hydrological Processes* **29** (7), 1873–1882.

Merz, R., Parajka, J. & Blöschl, G. 2011 Time stability of catchment model parameters: implications for climate impact analyses. *Water Resources Research* **47** (2). doi:10.1029/2010WR009505.

O'Kane, J. P. & Flynn, D. 2007 Thresholds, switches and hysteresis in hydrology from the pedon to the catchment scale: a non-linear systems theory. *Hydrology and Earth System Sciences* **11** (1), 443–459. https://doi.org/10.5194/hess-11-443-2007.

Page, T., Beven, K. J. & Freer, J. 2007 Modelling the chloride signal at the plynlimon catchments, wales using a modified dynamic TOPMODEL. *Hydrological Processes* **21**, 292–307.

Pappenberger, F., Matgen, P., Beven, K. J., Henry, J.-B., Pfister, L. & de Fraipont, P. 2006 Influence of uncertain boundary conditions and model structure on flood inundation predictions. *Advances in Water Resources* **29** (10), 1430–1449. doi:10.1016/j.advwatres.2005.11.012.

Parkin, G., O'Donnell, G., Ewen, J., Bathurst, J. C., O'Connell, P. E. & Lavabre, J. 1996 Validation of catchment models for predicting land-use and climate change impacts. 2. Case study for a Mediterranean catchment. *Journal of Hydrology* **175** (1–4), 595–613.

Peel, M. C. & Blöschl, G. 2011 Hydrological modelling in a changing world. *Progress in Physical Geography* **35** (2), 249–261.

Pool, S., Vis, M. J., Knight, R. R. & Seibert, J. 2017 Streamflow characteristics from modeled runoff time series importance of calibration criteria selection. *Hydrology and Earth System Sciences* **21** (11), 5443–5457.

Refsgaard, J. C. & Knudsen, J. 1996 Operational validation and intercomparison of different types of hydrological models. *Water Resources Research* **32** (7), 2189–2202.

Refsgaard, J. C., Storm, B. & Clausen, T. 2010 Système Hydrologique Europeén (SHE): review and perspectives after 30 years development in distributed physically-based hydrological modelling. *Hydrology Research* **41** (5), 355–377.

Reggiani, P. & Schellekens, J. 2003 Modelling of hydrological responses: the representative elementary watershed approach as an alternative blueprint for watershed modelling. *Hydrological Processes* **17** (18), 3785–3789.

Reggiani, P., Sivapalan, M. & Hassanizadeh, S. M. 2000 Conservation equations governing hillslope responses: exploring the physical basis of water balance. *Water Resources Research* **36** (7), 1845–1863.

Romanowicz, R. & Beven, K. J. 2003 Bayesian estimation of flood inundation probabilities as conditioned on event inundation maps. *Water Resources Research* **39** (3), W01073. doi:10.1029/2001WR001056.

Romanowicz, R., Young, P. C. & Beven, K. J. 2006 Data assimilation and adaptive forecasting of water levels in the River Severn catchment, UK. *Water Resources Research* **42**, W06407. doi:10.1029/2005WR004373.

Seibert, J. 2003 Reliability of model predictions outside calibration conditions. *Nordic Hydrology* **34**, 477–492. doi:10.2166/nh.2003.028.

Seibert, J. & McDonnell, J. J. 2002 On the dialog between experimentalist and modeler in catchment hydrology: use of soft data for multicriteria model calibration. *Water Resources Research* **38** (11), 23–21.

Seibert, J., Vis, M. J., Lewis, E. & Meerveld, H. V. 2018 Upper and lower benchmarks in hydrological modelling. *Hydrological Processes* **32** (8), 1120–1125.

Todini, E. & Mantovan, P. 2007 Comment on: 'on undermining the science?' by Keith Beven. *Hydrological Processes* **21** (12), 1633–1638.

Wallis, S. G., Young, P. C. & Beven, K. J. 1989 Experimental investigation of the aggregated dead zone model for longitudinal solute transport in stream channels. *Proceedings of the Institution of Civil Engineers (London)* **87**, 1–22.

Westerberg, I. K., Wagener, T., Coxon, G., McMillan, H. K., Castellarin, A., Montanari, A. & Freer, J. 2016 Uncertainty in hydrological signatures for gauged and ungauged catchments. *Water Resources Research* **52** (3), 1847–1865.

Westra, S., Thyer, M., Leonard, M., Kavetski, D. & Lambert, M. 2014 A strategy for diagnosing and interpreting hydrological model nonstationarity. *Water Resources Research* **50**, 1–24. doi:10.1002/2013WR014719.

Weyman, D. R. 1970 Throughflow on hillslopes and its relation to the stream hydrograph. *Hydrological Sciences Bulletin* **15**, 25–33.

Weyman, D. R. 1973 Measurements of the downslope flow of water in a soil. *Journal of Hydrology* **20** (3), 267–288.

Wilby, R. L., Clifford, N. J., De Luca, P., Harrigan, S. O., Hillier, J. K., Hodgkins, R., Johnson, M. F., Matthews, T. K. R., Murphy, C., Noone, S. J., Parry, S., Prudhomme, C., Rice, S. P., Slater, L. J., Smith, K. A. & Wood, P. J. 2017 The 'dirty dozen' of freshwater science: detecting then reconciling hydrological data biases and errors. *WIRES Water* **4**, e1209. doi:10.1002/wat2.1209.

Winsemius, H. C., Schaefli, B., Montanari, A. & Savenije, H. H. G. 2009 On the calibration of hydrological models in ungauged basins: a framework for integrating hard and soft hydrological information. *Water Resources Research* **45**, W12422.

Wood, E. F., Roundy, J. K., Troy, T. J., Van Beek, L. P. H., Bierkens, M. F., Blyth, E., de Roo, A., Döll, P., Ek, M., Famiglietti, J. & Gochis, D. 2011 Hyperresolution global land surface modeling: meeting a grand challenge for monitoring Earth's terrestrial water. *Water Resources Research* **47** (5), W05301. doi:10.1029/2010WR010090.

Young, P. 1998 Data-based mechanistic modelling of environmental, ecological, economic and engineering systems. *Environmental Modelling & Software* **13** (2), 105–122.

Young, P. C. & Beven, K. J. 1994 Data-based mechanistic modelling and the rainfall-flow non-linearity. *Environmetrics* **5**, 335–363.

Zehe, E. & Jackisch, C. 2016 A Lagrangian model for soil water dynamics during rainfall-driven conditions. *Hydrology and Earth System Sciences* **20**, 3511–3526. https://doi.org/10.5194/hess-20-3511-2016.

Zreda, M., Shuttleworth, W. J., Zeng, X., Zweck, C., Desilets, D., Franz, T. & Rosolem, R. 2012 COSMOS: the COsmic-ray Soil Moisture Observing System. *Hydrology and Earth System Sciences* **16**, 4079–4099. https://doi.org/10.5194/hess-16-4079-2012.

First received 10 October 2018; accepted in revised form 13 January 2019. Available online 23 October 2019

Proposal of the 'extreme rank plot' for extreme value analysis: with an emphasis on flood frequency studies

Anthony Hammond

ABSTRACT

To assess model 'fit' for flood frequency studies, the annual maximum flow sample is often plotted alongside the frequency curve in an extreme value plot. The comparison is ill advised because a value is speculatively assigned for each rank, in the form of a frequency (the plotting positions); in which case a model of the frequencies is being used to assess a frequency model. The modelled and observed, with attributed plotting positions, are incommensurable. The proposed method in this paper does not attempt to provide a frequency attributed to the ranks and instead attributes ranks to the model under investigation. If the model is used to simulate thousands of samples of the same size as the observed, a distribution of possible peak flow values for each rank can be obtained. For each simulated sample, descriptive statistics can be derived. In this way, statistical testing can be undertaken to compare the observed with the modelled. Using such methods, two or more models can also be compared. Examples are provided for Flood Estimation Handbook pooling and continuous simulation models along with a comparison of the two.

Key words | continuous simulation, extreme rank plot, extreme value analysis, flood frequency estimation, plotting positions

Anthony Hammond
JBA Risk Management,
1 Broughton Park, Old Lane,
Broughton, Skipton,
UK
E-mail: *anthony.hammond@jbarisk.com*

INTRODUCTION

In flood frequency studies, the frequency plot or the extreme value plot, as it is also known (henceforth named the EV plot), is ubiquitous. It is recommended in the Flood Estimation Handbook (FEH; Robson & Reed 1999), and no update since has suggested a diversion from its use. However, in the author's experience of many industry-based studies, the EV plot appears to have little to no impact on the outcome of the decisions made in regards to the validation of the model. Where the observed appears to fit the frequency curve well, it is seen as promising and validation of the method. Where it does not, it is often stated that the EV plot is misleading because the plotting positions provide a false representation of the frequency of the observed flows. With these two outcomes, the EV plot provides very little benefit. The EV plot used in this manner could be detrimental because using it, and dismissing its results, appears to preclude further investigation. The root of the problem is the plotting positions. In flood frequency analysis, the objective is to determine flood frequencies; however, to compare the frequency model with the observed data using the EV plot, the frequencies must first be estimated. The plotting positions facilitate this purpose and are themselves a model of the frequencies. The EV plot, therefore, compares one model of the frequency with another. A suggested solution is to invert the problem and attribute ranks to the model instead. This approach is outlined below with an FEH-based example and includes a hypothesis test which is facilitated by the proposed solution. An example using continuous simulation (CS), with a comparison between the FEH and CS results, is

doi: 10.2166/nh.2019.157

provided in the third section. Firstly, a summary of the EV plot, plotting positions and the inherent problems is provided, but not before a note about the data used in the study.

Data

For the examples, the recorded data used in this study is from the UK's National River Flow Archive (NRFA): NRFA peak flow dataset – version 7. The CS data is from a study undertaken by Jeremy Benn Associates on behalf of the Environment Agency and is discussed in more detail for the relevant example. All the data analysis and associated plotting were undertaken using base 'R' (R Core Team 2014).

A BRIEF SUMMARY OF THE EV PLOT

To judge the accuracy of a model, it is necessary to consider its accuracy in terms of its agreement with the data used to estimate its parameters or its agreement with a sample of data from the population it is attempting to model (often one and the same). A visual comparison can quickly highlight any significant lack of agreement, and this is the primary purpose of the EV plot. The EV plot, or any plot of ordered data using plotting positions, aims to inspect the distribution of the ordered data. The plotting positions provide an estimate of the cumulative frequency for the ith term (Gringorten 1963), in which case the plot has frequency on the abscissa and magnitude on the ordinate. Commonly, the frequency scale is further adjusted as a reduced variate, which has the benefit of providing a more linear plot and putting more emphasis on the rarer events. It also has the benefit of illustrating whether the distribution being plotted is bounded above or not (Robson & Reed 1999). However, the reduced variate has the disadvantage of relatively meaningless labels on the abscissa, requiring the plot of an additional horizontal axis to provide a more meaningful measure of frequency. For example, where T is return period, Robson & Reed (1999) suggest a logistic reduced variate $y_l = \ln(T-1)$, which provides a range of -4 to 4 over the return periods from 1.0183 to 55 (approximately). The secondary axis then plots the return periods at the relevant location of $\ln(T-1)$. In UK flood frequency analysis, the Gringorten formula is generally recommended (Shaw 1993; Robson & Reed 1999) and provides the following frequencies, from which T, in the reduced variate scale, is derived:

$$F_i = i^{\text{th}} \text{ plotting position} = \frac{i - 0.44}{n + 0.12} \quad (1)$$

where n is the sample size. This plotting position was outlined by Gringorten (1963) for use with the double exponential distribution, otherwise known as the Gumbel distribution or extreme value type 1 distribution. This secondary axis provides clarity above the reduced variate alone. In practice, it is not of particular use for determining discharge for a given return period, or vice versa; where tabulated results are preferable if precision is required. The commonly used software used to undertake the FEH statistical methods is WINFAP (Wallingford Hydro Solutions), and it generates the EV plot using the method outlined by Robson & Reed (1999). Figure 1 shows an example for NRFA site 39001, formatted in a similar way to the WINFAP-generated plots.

Plotting positions

As noted, to determine the frequency for the abscissa, plotting positions are required and provide frequency as a

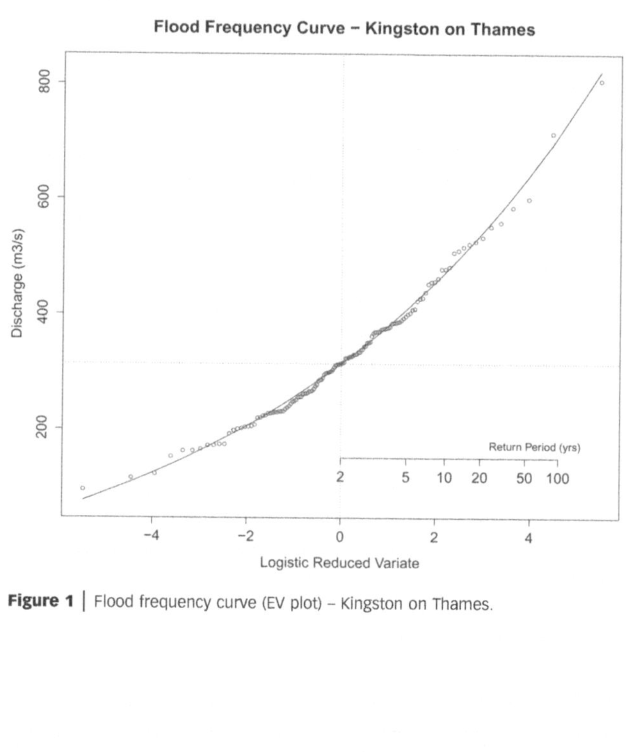

Figure 1 | Flood frequency curve (EV plot) – Kingston on Thames.

function of the ordered ranks. Gringorten (1963) and Sooyoung *et al.* (2012) discuss the development of plotting positions for given distributions, whereas others promote distribution-free approaches (Gumbel 1958; Cunnane 1978; Makkonen 2008). The debate about the best approach is evident when considering the title of Makkonen's (2008) paper; 'Bringing Closure to the Plotting Position Controversy'. The last sentence of the abstract reads: 'The article is intended to mark the end of the century-long controversial discussion on the plotting positions'. In that paper, it is argued that the distribution-free Weibull formula is used, as suggested by Gumbel (1958):

$$F_i = \frac{i}{n+1} \tag{2}$$

The justification for a distribution-free plotting position, such as Equation (2), is simple enough; the plotting position should not be considered an estimate itself and therefore should be chosen independently of distribution. The proposal in this paper takes this justification a step further and to a logical conclusion; no estimated frequencies–no plotting positions. The Weibull formula is independent of the distribution being considered but is still an estimate of the ith ordered frequency. With a large enough sample, the Weibull formula should, as Makkonen (2008) argues, provide the probability of non-exceedance of the ith value in n-ranked events. However, Cunnane (1978), who provides a review of 'unbiased' plotting positions, shows that the Weibull is biased for all but the uniform distribution. As noted, in the same paper, the unbiased plotting positions are the mean of the ith order statistic from the reduced variate sampling distribution, which differs from distribution to distribution. Therefore, a single choice of plotting position will have varying levels of bias across the range of distributions. To test, empirically, for the bias of a plotting position for the ith flow in an ordered sample, the following procedure can be undertaken:

1. Using the quantile function of the distribution being considered, simulate and order 10,000 samples of size n.
2. Attribute a frequency to the ith flow with the plotting position formula under investigation.
3. Use the same quantile function used in step 1 to generate a flow as a function of the attributed frequency.
4. Compare the mean of the ith distribution of flows with the result from step 3.
5. Repeat for numerous sample sizes.

With the sample of annual maxima (AMAX) from the Kingston gauge (site 39001), this procedure, for the largest flow, was completed for the Weibull and the Gringorten plotting position formulas, with both the generalised logistic and generalised extreme value (GEV) distributions. Table 1 shows the results for sample sizes 5, 10, 20, 50, 100 and 200.

As can be seen in Table 1, bias is evident. A further point is the inherent uncertainty, as noted by Cook (2012) 'uncertainty due to sampling error dominates over the choice of methodology'.

This is aptly demonstrated by Figures 2 and 3 and Table 2, which all summarise the following: using the GEV distribution, 10,000 samples of size 42 were simulated using UK average parameters from the sites in the NRFA data set considered suitable for pooling. Every sample was ordered to provide a distribution of 10,000 possible results for each rank. The variance of the distribution of possible flows over a given rank increases with magnitude.

The mean inter quartile range (IQR) for the simulated AMAX is 38.9 m^3/s, which is similar to the IQR of the simulated 42nd rank. The possible IQR for the maximum rank distribution being slightly greater than the likely IQR of the AMAX sample suggests that frequencies attributed to the rank by plotting positions alone cannot be usefully informative when assessing model fit. As noted by Stedinger *et al.* (1993), differences between plotting positions are modest for $i = 3$ or more, but differences can be appreciable for the largest observations. Stedinger *et al.* (1993) also note that a good method for illustrating the uncertainty is to consider the distribution of the largest observation to be beta-distributed. In this case, the actual exceedance probability in a sample is between $0.052/n$ and $3/(n+2)$ nearly 90% of the time. Such intervals could be added to the EV plot, as could bootstrapped confidence intervals. However, in the former case, the intervals could be either side of a biased central estimate and in the latter case, the bootstrapped intervals would show the same bias as the central estimate. Little is gained from assuming the beta distribution to illustrate uncertainty in place of simulation approaches, which are quick and simple.

Table 1 | Gringorten and Weibull plotting position – bias test results

Gringorten			GLO			GEV		
Sample size	Rank 1 frequency	*T*	Mean	Rank 1 flow	Difference	Mean	Rank 1 flow	Difference
5	0.1094	9	472.6	461.4	11.2	469.1	471.5	−2.5
10	0.0553	18	536.4	525.4	11.0	530.5	531.7	−1.2
20	0.0278	36	607.2	593.9	13.4	586.1	589.1	−3.0
50	0.0112	90	709.3	693.5	15.8	659.8	661.4	−1.6
100	0.0056	179	791.8	777.1	14.7	710.8	713.8	−3.0
200	0.0028	357	888.7	868.8	19.9	760.8	764.3	−3.5
Weibull			**GLO**			**GEV**		
Sample size	**Rank 1 frequency**	***T***	**Mean**	**Rank 1 flow**	**Difference**	**Mean**	**Rank 1 flow**	**Difference**
5	0.1667	6	472.6	422.6	50.0	469.1	432.0	37.1
10	0.0909	11	536.4	478.5	58.0	530.5	488.3	42.2
20	0.0476	21	607.2	540.0	67.3	586.1	544.5	41.6
50	0.0196	51	709.3	630.8	78.5	659.8	617.3	42.5
100	0.0099	101	791.8	707.5	84.3	710.8	670.7	40.1
200	0.0050	201	888.7	792.0	96.8	760.8	722.5	38.3

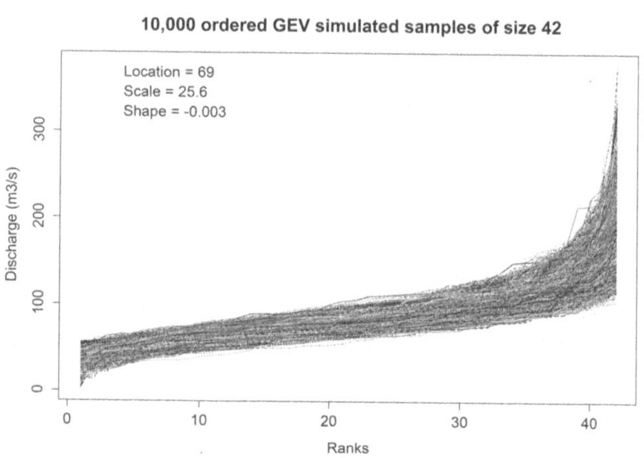

Figure 2 | Ordered GEV-simulated samples demonstrating a range of possible flows at each rank.

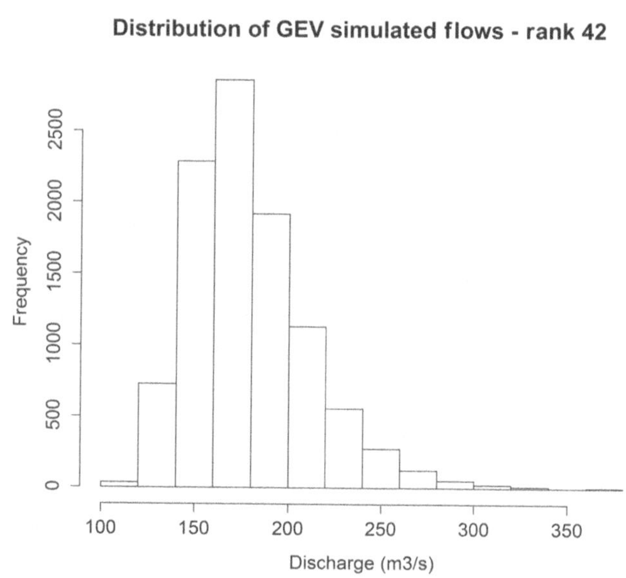

Figure 3 | Distribution of flows at rank 42 of the GEV-simulated AMAX.

The extreme rank plot

To address the issue of plotting positions, the problem can be inverted. The ranks of the observed data are known, not the frequencies. Rather than attributing the frequencies to the observed data, ranks can be attributed to the modelled distribution. The ranks are not known for the modelled, hence an inverted problem. Fortuitously, individual expected ranks for the modelled distribution are not necessary for comparison, only a distribution of probable flows for each rank. Using the model under consideration, if j samples of observed sample size n are simulated, and ordered, a distribution of flows of size j can be determined for each rank. The observed data can then be ordered and plotted with the distribution of flows for

Table 2 | Distribution statistics for ranks of the GEV-simulated AMAX

Rank	Minimum	First quantile	Median	Mean	Third quantile	Maximum	IQR	0.025th	0.975th	95% range
10	42.3	55.8	59.1	59.1	62.4	79.1	6.6	50	69	19
20	56.8	72.2	75.8	76.0	79.6	98.7	7.4	66	87	21
30	71.4	90.7	95.4	95.7	100.4	130.3	9.7	83	111	28
40	94.1	128.7	138.3	139.9	149.3	237.0	20.7	113	176	63
42	108.2	155.6	173.3	178.4	195.0	389.5	39.4	131	257	126

each rank, providing an extreme rank plot (ER plot). The way in which the observed AMAX flows fit within the modelled distribution of flows at each rank provides an indication of how well the model represents the underlying distribution from which the observed AMAX originates. To summarise, the process for developing the ER plot for a given frequency model and observed data is as follows:

1. Using the parameter estimates and the quantile function of the frequency model, simulate AMAX (10,000 (for example) multiplied by the observed sample size).
2. Split the results from step 1 into 10,000 separate samples, the same size as the observed and sort them into ascending order.
3. For each rank in the ordered data, calculate upper and lower and central nonparametric quantiles of the 10,000 simulated flows (for example, the 0.025th, 0.5th, and 0.975th quantiles).
4. Sort the observed AMAX series into ascending order and plot (abscissa = rank, ordinate = flow) with the simulated quantiles for each rank.

ER PLOT EXAMPLE FOR NRFA SITE 39056

The ER plot could be used for any frequency model; in this case as an example, flood frequency analysis was undertaken for site 39056 in the NRFA data set using the FEH pooling model (Kjeldsen *et al.* 2008). Firstly, a brief summary of the FEH pooling approach is outlined.

For a sequence of daily peak flows ($x1$, $x2$ ……, $x365$), there exists:

$$Q = \max\{x1 \ldots\ldots x365\} \tag{3}$$

which is the annual maximum peak flow. A sample of Q from a gauged record is the annual maximum sequence (AMAX) for which a probability distribution can be assumed. If x is flood magnitude, then the probability that an annual maximum equals or exceeds x is:

$$P(x) = \int_x^\infty p(Q)dQ \tag{4}$$

The reciprocal of Equation (4) is the return period (T). The FEH default recommendation for distribution to model the AMAX is the generalised logistic distribution (GLO), for which flow as a function of T is:

$$Q_T = \mu + \frac{\alpha}{k}(1 - (T - 1)^{-k}) \tag{5}$$

where μ, α and k are location, scale and shape parameters of the GLO distribution, respectively, and can be estimated by the method of L-moments (Hosking & Wallis 1997). Where there is insufficient gauged flow data to estimate the parameters at the site of interest, the FEH recommends regional flood frequency analysis, using an index flood procedure, specifically utilising the L-moment method. The region (or pooled group) is a set of gauged sites with catchment descriptors that are deemed sufficiently similar to those of the site of interest, measured by a similarity distance measure (SDM). The assumption is that sites in the pooling group have the same underlying AMAX distribution, except for a scaling factor (the index flood). The quantile function using the index flood procedure is:

$$Q_T = \mu\left[1 + \frac{\beta}{k}(1 - (T - 1)^{-k})\right] = \mu\, q(T) \tag{6}$$

The term between the squared brackets is a dimensionless growth factor ($q(T)$) and is the pooled growth curve. The index flood (μ) is the median annual maximum flow (QMED) and is used as the location parameter. The β and k parameters used to form the dimensionless growth curve are estimated using L-moment ratios, LCV (t_2) and LSkew (t_3), where $k = -t_3$ and

$$\beta = t_2 \frac{k \sin \pi k}{k\pi(k + t_2) - t_2 \sin \pi k} \quad (7)$$

A weighted average of L-moment ratios from the pooling group is used for the final estimation of the dimensionless growth curve for the site of interest. The index flood is derived as the median of the AMAX sample at the site of interest, or where there is no data, by the regression model with catchment descriptors as the explanatory variables.

Finally, for the purpose of the following example, it should be noted that the majority of catchments in the NRFA are classified as rural. In which case, the majority of sites in any given pooling group would be rural unless a significant reduction of candidate sites is relied upon. For this reason, urban sites are not included in pooling groups, including their own pooling group. The allowance for catchment urbanisation is considered separately with an adjustment. The urbanisation is described by the proportion of urban and suburban land cover within the catchment, which is summarised by the statistic URBEXT.

Details of the formation of pooling groups, L-moment ratios, adjusting for urbanisation and the estimation of the index flood are provided by Kjeldsen *et al.* (2008) and Kjeldsen (2010).

The example site 39056 is on the River Ravensbourne at Catford and is summarised with catchment descriptors in Table 3 and with the annual maximum sample in Figure 4. Table 4 shows the pooling group, and resulting weighted mean LCV (linear coefficient of variation) and LSkew (L-moment 3 divided by L-moment 2) (Hosking & Wallis 1997) are summarised in Table 5.

Table 5 shows the resulting L-moment ratios from the gauged data, the weighted ratios from the as rural pooled group and the ratios after an urban adjustment.

Table 3 | Relevant catchment descriptors for NRFA site 39056

Descriptor	Details	Value
AREA	Catchment drainage area in km^2	126.18
SAAR6190	Standard annual average rainfall from 1961 to 1990 (mm)	714
FARL	Index of flood attenuation from reservoirs and lakes	0.99
BFIHOST	Base flow index derived from hydrology of soil classification	0.715
FPEXT	Proportion of catchment in the 1 in 100 year floodplain	0.0613
URBEXT2000	Extent of urban and suburban land cover (2000)	0.3429

Using the resulting LCV and LSkew, the parameters of the chosen distribution can be determined to simulate multiple samples the same size as observed. In the pooling case, an estimate of QMED is multiplied by the growth factor to form the quantile function. For this example, the GLO was used as outlined in the FEH summary (Equation (7)).

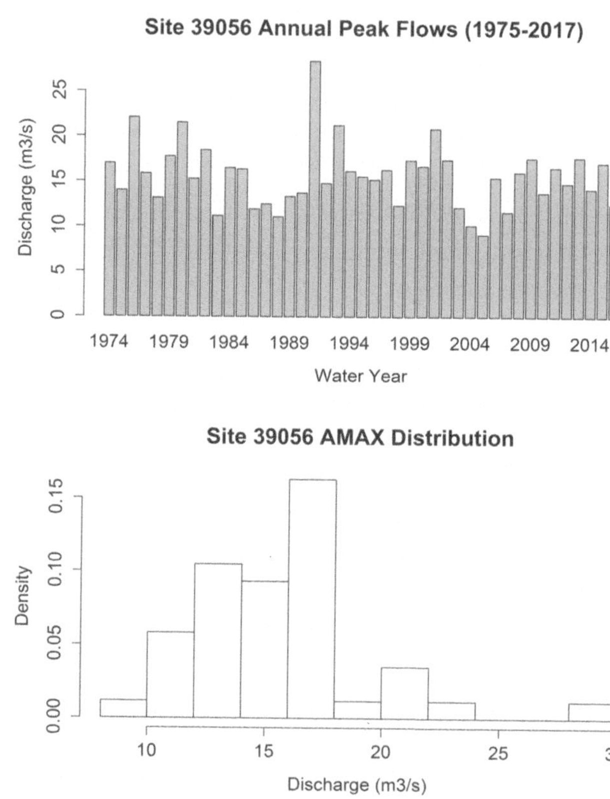

Figure 4 | Summary plots for the annual maximum sample at NRFA site 39056.

Table 4 | FEH pooling group for NRFA site 39056

Site	SDM	*N*	QMED	LCV	LSkew
33018 (Tove @ Cappenham Bridge)	0.165	54	16.95	0.285	0.173
38004 (Rib @ Wadesmill)	0.296	58	11.718	0.318	0.155
21016 (Eye Water @ Eyemouth Mill)	0.305	39	36.964	0.275	0.151
39025 (Enborne @ Brimpton)	0.307	50	17.139	0.198	0.148
21027 (Blackadder Water @ Mouth Bridge)	0.346	32	40.298	0.321	0.268
33051 (Cam @ Chesterford)	0.381	48	8.017	0.245	−0.108
37020 (Chelmer @ Felsted)	0.386	42	13.44	0.284	0.249
39028 (Dun @ Hungerford)	0.394	49	2.19	0.229	−0.01
20003 (Tyne @ Spilmersford)	0.4	56	34.78	0.372	0.215
43004 (Bourne @ Laverstock)	0.411	45	2.157	0.328	0.31
13001 (Bervie @ Inverbervie)	0.425	27	35.577	0.212	0.141
Weighted means				0.279	0.153

Table 5 | LCV and LSkew results for site 39056

Statistics	Observed	Pooled rural	Urban adjusted
LCV	0.124	0.279	0.228
LSkew	0.115	0.153	0.211

Therefore, in the case of the pooling group for site 39056,

$$\beta = \frac{0.279\{-0.153\sin[\pi(-0.153)]\}}{-0.153\pi(-0.153+0.279)-0.279\sin[\pi(-0.153)]} = 0.288 \quad (8)$$

and

$$Q_T = 15.546\left[1+\frac{0.288}{-0.153}(1-(T-1)^{0.211})\right] \quad (9)$$

Using Equation (9), 10,000 times n years of AMAX (43) were simulated, split into 10,000 samples of size n and ordered. This provided a distribution of 10,000 flow estimates for each rank. The 0.025th, 0.5th and 0.975th nonparametric quantiles for each rank were determined using quantile definition seven outlined by Hyndman & Fan (1996), which is the default function in commonly used statistical software packages. These were then plotted alongside the ordered observed data. Figure 5 is the resulting ER plot.

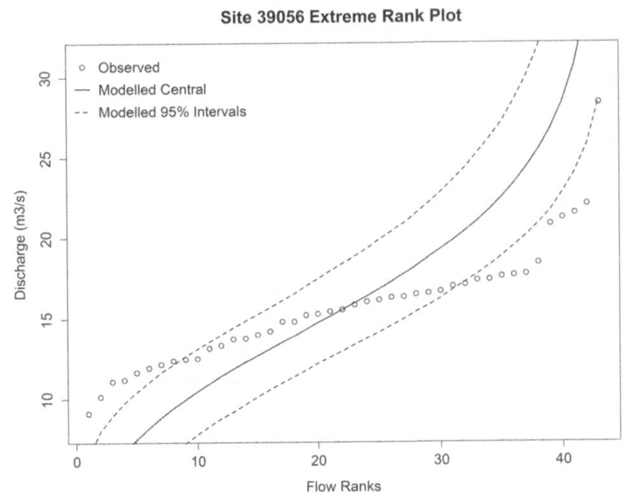

Figure 5 | ER plot for NRFA site 39056 (as rural pooling model).

As site 39056 has an URBEXT2000 value of 0.3429, the site is considered as heavily urbanised. Packman (1980) suggested that urbanisation would impact the scale of the distribution (and therefore the LCV), which was supported by the findings of Kjeldsen (2010). This may account for the significant difference in the pooled as rural LCV and the observed. With the urban adjustment suggested by Kjeldsen (2010), the LCV and LSkew become 0.228 and 0.211, respectively (as shown in Table 5). This provides the urban-adjusted version of the ER plot in Figure 6. Figure 7

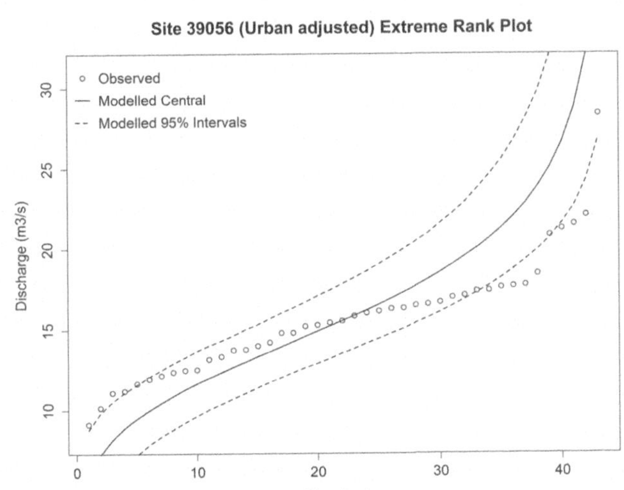

Figure 6 | ER plot for NRFA site 39056 (pooling model with urban adjustment).

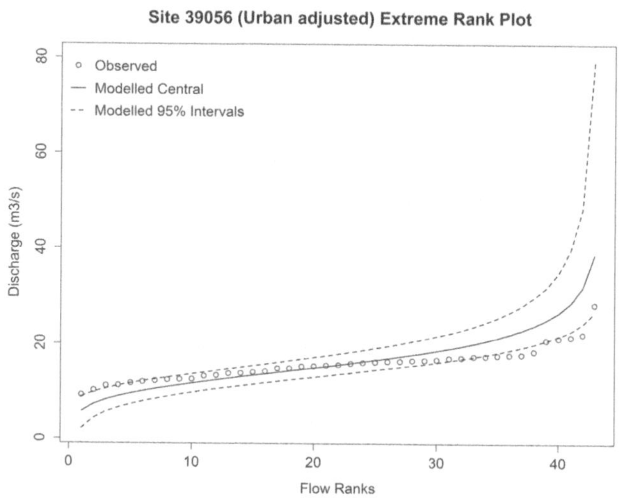

Figure 7 | Repetition of Figure 6 zoomed out for full view of intervals.

provides the same but is zoomed out to show the full extent of the 95% intervals, and Figure 8 provides, for comparison, the EV plot with urban adjustment for the same site.

As can be seen from the ER plot and the EV plot, the ungauged pooled estimate does not appear to fit the observed distribution well and is marginally improved by the urban adjustment. As a final comparison between the ER plot and EV plot, Figure 9 provides both with parameters estimated from the at site AMAX sample, for the longest record in the NRFA dataset; site 39001.

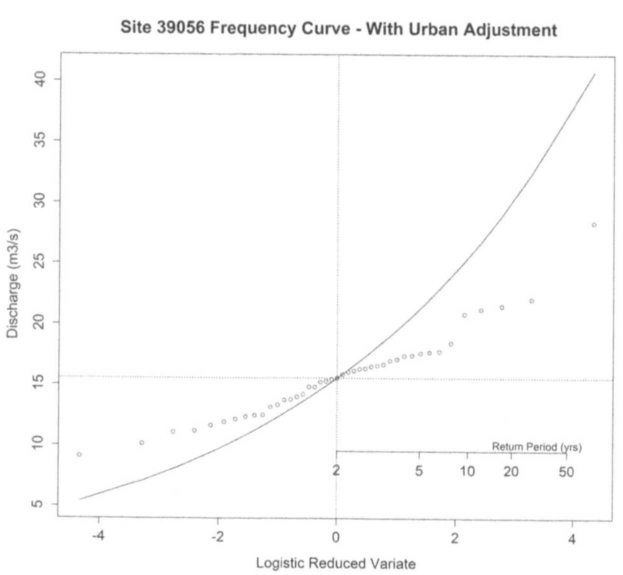

Figure 8 | EV plot with a pooled frequency curve with urban adjustment for NRFA site 39056.

Hypothesis testing

The ER plot can be strengthened quantitatively by a hypothesis test using the simulated data. If LCV and LSkew statistics are derived for each simulated AMAX, the distribution of these statistics can be compared to the LCV and/or LSkew of the observed data. The hypothesis test can be stated as follows:

H_0 = the observed AMAX is a sample from the modelled distribution

H_1 = the observed AMAX is not a sample from the modelled distribution

For the purposes of this paper, setting an arbitrary alpha level (significance level) was not considered necessary.

In the case of site 39056, the LCV and LSkew are 0.124 and 0.115, respectively.

Table 6 and Figure 10 show the distributions of LCVs and LSkews from the simulated samples using the urban-adjusted LCV and LSkew in Table 5.

Using the simulated statistics, p-values can be determined and provide the probability of observing the statistics if the observed AMAX did originate from the estimated distribution. Taking the ratio of simulated statistics above that of the observed to the total number of simulated statistics:

$$p\text{-value}(S) = \begin{cases} \dfrac{N_{\text{Ssim}} > S_{\text{obs}}}{N_{\text{Ssim}}}, & (N_{\text{Ssim}} > S_{\text{obs}}) < \dfrac{N_{\text{Ssim}}}{2} \\ 1 - \left(\dfrac{N_{\text{Ssim}} > S_{\text{obs}}}{N_{\text{Ssim}}}\right), & (N_{\text{Ssim}} > S_{\text{obs}}) > \dfrac{N_{\text{Ssim}}}{2} \end{cases} \tag{10}$$

where S is the descriptive statistic under consideration, N_{Ssim} is the number of statistics derived from the simulated data and S_{obs} is the statistic from the observed sample.

The pooling with an urban adjustment model for site 39056 provides an LCV less than the minimum simulated, suggesting a p-value $<1\times10^{-4}$. That is, there is less than a 0.0001 probability (<1 in 10,000) of observing the AMAX sample if it did originate from the modelled distribution. This provides strong evidence for the rejection of the null hypothesis (H_0). In which case, there are one of two conclusions,

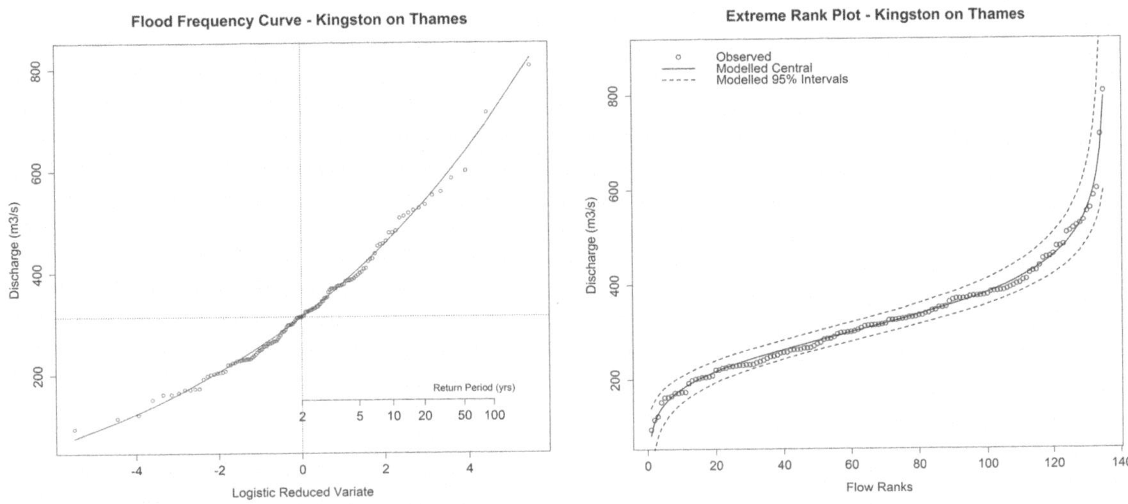

Figure 9 | EV plot and ER plot for NRFA site 39001 (single-site analysis).

Table 6 | Statistics for the simulated LCVs and LSkews

Statistic	Minimum	First quantile	Median	Mean	Third quantile	Maximum	SD
LSkew	−0.178	0.081	0.143	0.144	0.204	0.5037	0.092
LCV	0.134	0.2046	0.224	0.227	0.2455	0.465	0.036

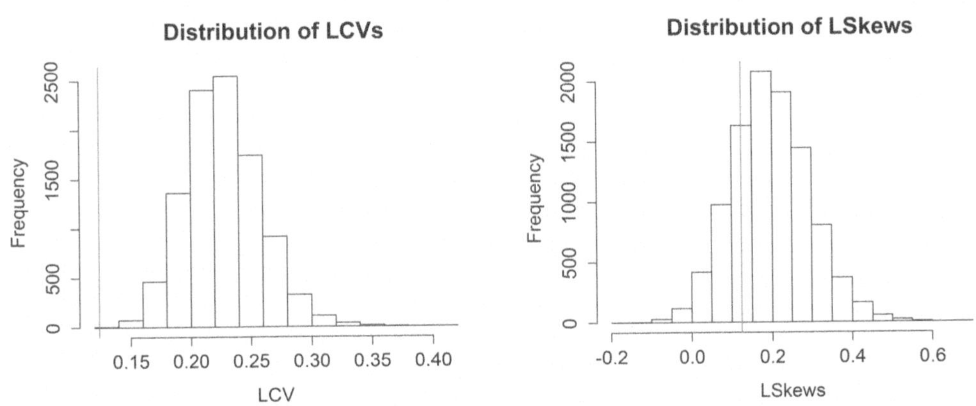

Figure 10 | Simulated distributions of LCV and LSKEW statistics. The vertical line places the observed LCV and LSkew on the distributions.

either the flow gauging is in error or the modelled frequencies are not adequate. Either way, more work is needed, such as a check of the gauging and further consideration for the frequency analysis. In the case of site 39056, the author reviewed a flood frequency analysis report, which accepted the FEH pooling model on the grounds that the plotting positions were not representative of the observed AMAX frequencies. The associated river levels attributed to the estimated flood frequencies were being used to justify and design a flood alleviation scheme. Therefore, due to a single plotting method or at least uncertainty about the adequacy of the plotting method, inadequate or unnecessary schemes could be developed. Without the plotting positions and with use of the ER plot and associated hypothesis test, a reconsideration of the gauged data and the frequency estimation would have been a more likely outcome.

It is notable that this test statistic approach shares similarities with the goodness-of-fit (GOF) measure outlined by

Hosking & Wallis (1997) and further developed by Kjeldsen & Prosdocimi (2015). However, whereas these GOF measures are specific to assessing the fit of distributions for regional data with L-moments, the hypothesis test outlined is broader and can only be used when there is at site data. The two are also not mutually exclusive because, in the case of the FEH approach, the final model which was, in part, formed by the GOF measure is then what is applied to the ER plot and the associated hypothesis test. The generality of the approach means it can be used to assess any model that can simulate data and be compared to observed data; any statistic of interest can be considered.

In cases where the LCV or LSkew are specifically used to form the model, such as AMAX single-site analysis (fitting distributions with L-moments) or where the site is included in the pooling group, a different statistic is necessary for the development of a hypothesis test. For comparison of extreme value models, it seems appropriate that some measure of symmetry is considered because the shape has the largest impact on the extremes of interest. For this purpose and for examples in the following section, two measures of skewness were considered, the skewness coefficient and the Yule Kendall index. The first is considered to be neither robust nor resistant enough (Wilks 2011). The second does not make use of the larger flows and, although robust and resistant, lacks power. For example, the distribution at Kingston (NRFA site 39001) is clearly right skewed but the resulting Yule Kendall index, using only the central 50% of data, does not reflect this. The ratio of the median to the mean of the data is considered to be a good compromise as a skew statistic – the median mean ratio (MMR). When there is a right skew, the MMR is positive, and when there is a left skew, the MMR is negative. If q is quantile, then:

$$\mathrm{MMR} = 1 - \frac{q_{0.5}(x)}{n^{-1}\sum_{i=1}^{n} x_i} \tag{11}$$

Use with CS and comparing CS with FEH

As detailed in the FEH summary, future flood events are estimated, using a probability function, as the probability of a given peak flow being equalled or exceeded. With a large enough sample, it is not necessary to specify a probability function because the statistics of interest can be estimated empirically. Boughton & Droop (2003) define CS in the context of flood hydrology as 'the estimation of losses from rainfall and the generation of stream flow by simulating the wetting and drying of a catchment at daily, hourly and occasionally sub-hourly time steps'. If enough stream flow is generated using CS and an annual maximum series is extracted, Q_T can be estimated as the ith peak flow in the ordered sample if $i/n = p(x)$ in Equation (4), where n is the sample size of simulated AMAX. The CS approach to flood frequency analysis, therefore, consists of a long sequence of simulated rainfall, as the input for a rainfall runoff model, which generates a sufficiently long sequence of stream flow in order to estimate Q_T empirically.

The ER plot and hypothesis test, thus far outlined, could be adopted for CS modelling, except there is a finite modelled AMAX. In which case, random sampling with replacement (resampling) from the simulated AMAX can be undertaken to build the modelled distributions of the observed size. Aside from developing the numerous simulated distributions, the process can be applied in the same way.

With resampling, the hypothesis testing and plot can be used to compare CS estimated frequencies with that of other methods such as the FEH method. In which case, it is necessary to simulate the same number of AMAX samples from the FEH-derived distribution as were simulated with CS. The resampling and hypothesis tests can be undertaken on both and the associated p-values can be directly compared; the one with the higher p-value provides the better fit.

ER plot example for CS

In 2017, on behalf of the Environment Agency, Jeremy Benn Associates undertook CS modelling to determine flood frequencies for the rivers Darent and Cray (Wass 2017). NRFA site 40012 on the Darent at Hawley was assessed as part of the study, and the largest flow peak observed in the AMAX was reassessed and reduced from 49 to 26 m^3/s. Five thousand years of simulated rainfall was run through PDM rainfall runoff models (Moore & Clarke 1981), which were combined to represent the catchment as a set of disparate sub catchments. To save on simulation time, only events where at least one of the PDMs produced a

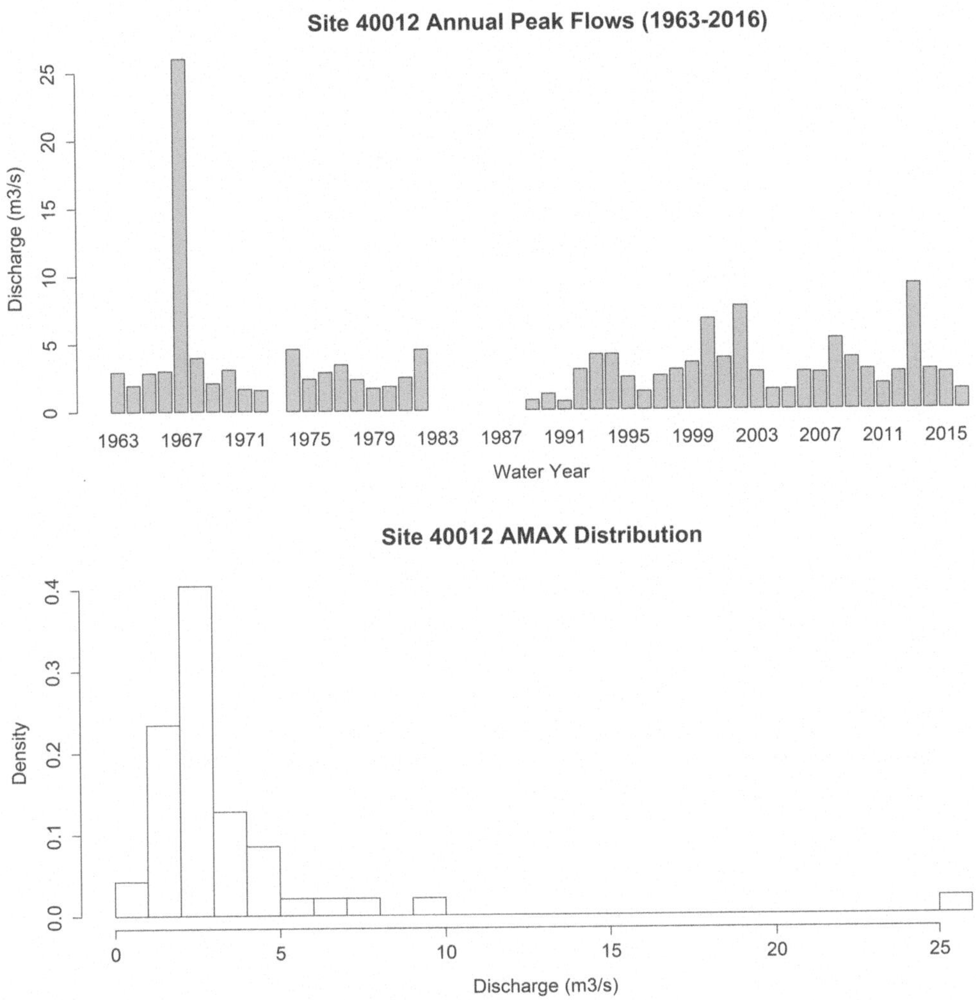

Figure 11 | Summary plots for the AMAX sample at NRFA site 40012.

flow above QMED were routed through to create the combined simulated flow for the catchment, which provided a total AMAX of 4080. Figure 11 provides summary plots of the AMAX sample at site 40012.

The ER plot in Figure 12 is the result of the resampling (size 1,000n) approach used with the 4,080 peaks from the JBA study. The 4080th peak is approximately the 0.184th percentile of the 5,000 year distribution. For this reason and for a valid comparison within the plot, the AMAX was also truncated at the 0.184th percentile leaving a sample of the 38 largest flows of the original 47.

Equation (5), with parameters; median (loc) = 2.74, beta = 0.332 and $k = -0.494$ (single-site parameter estimates), was used to simulate 5,000 annual maximum

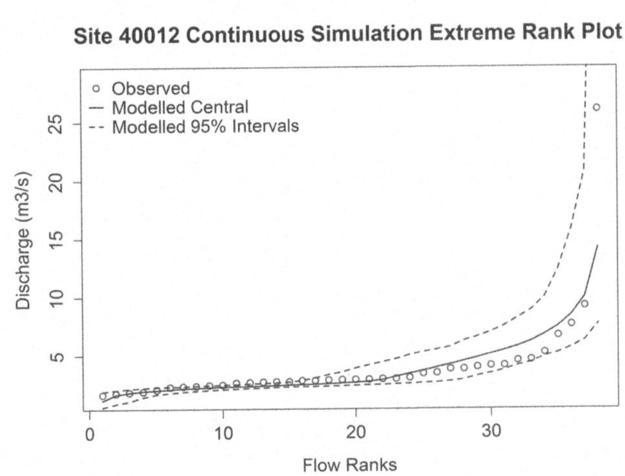

Figure 12 | ER plot for the CS model at NRFA site 40012.

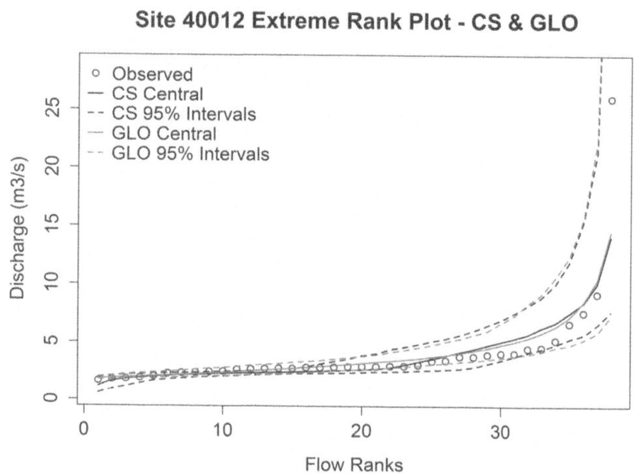

Figure 13 | ER plot for NRFA site 40012 with the CS model and the GLO model.

flows. For the necessary resampling (size 1,000n), this 5,000 was truncated at the 4,080th ranked flow. Figure 13 shows the ER plot with both the CS model and the GLO model included for comparison.

Equation (11) was used to calculate MMRs from the distributions formed from the resampling of both the CS and GLO models. Figure 14 compares the distributions with boxplots and includes a horizontal line to show the MMR of the truncated observed data at site 40012.

In this example, the p-values were estimated, using Equation (10), as 0.1 and 0.4 for the CS and the GLO models, respectively. Based on these p-values, both models appear to represent the distribution well, but the GLO performs slightly better. The resulting 1 in 100 flow estimate is 18 m^3/s for both.

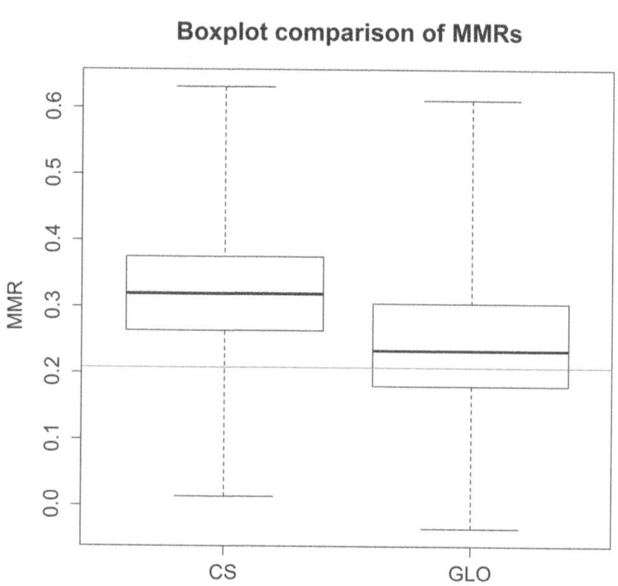

Figure 14 | Boxplot comparison of the MMR statistic for the 1000 resamples – CS and GLO. The horizontal line across both plots is the observed MMR at site 40012.

CONCLUSION

This paper introduced the ER plot for use in flood frequency analysis, in place of the EV plot, often known as the flood frequency curve plot. The incommensurable nature of the plotted variables in the EV plot was illustrated as was the manner in which the EV plot is misused due to a lack of confidence and/or understanding in the plotting positions. The proposed approach in this paper uses the model under consideration to model the ranks stochastically and compares central, upper and lower bounds of the stochastic output with the observed data. This was achieved by modelling 10,000n years of annual maximum events, splitting them up into 10,000 samples of size n and ordering them, providing a 10,000 AMAX distribution of modelled flows for each rank.

The 10,000 simulated samples were used to derive a distribution of statistics for comparison with the observed statistics for hypothesis testing. Where the observed data was not used to derive the LCV and LSkew for the model, these were used for the hypothesis test and associated p-values were determined to provide the probability of observing the data, had it originated from the modelled distribution.

An example was also provided for a CS study, the results of which were compared to the results of fitting a GLO directly to the AMAX data. In this latter case, the observed data had been used directly as an estimate of the LCV and LSkew for the model. Therefore, the MMR was used as a measure of skew for hypothesis testing. Due to the finite size of the simulation in the CS approach, resampling was used to build the distribution of flows for each rank (1,000 per rank). For comparison of the CS model with the GLO, the GLO was first used to simulate the same finite sample size as produced by the CS and then resampled in the same way.

The plot and the hypothesis testing made possible by the associated simulations are considered to be useful tools for flood frequency model validation.

It is recommended that the ER plot, along with the associated quantification of model fit and model comparison, be adopted for flood frequency studies. It is also recommended that further work is undertaken to determine the most useful statistic for use in the hypothesis testing and comparing of models. It is envisaged that 'most useful' in this case would be a statistic that has the greatest impact on the modelled events of interest.

ACKNOWLEDGEMENTS

The author would like to thank all those responsible for the availability of the National River Flow Archive and the British Hydrological Society for the opportunity to publish this study. The author would also like to thank the anonymous reviewers for their helpful comments.

REFERENCES

Cook, J. 2012 Rebuttal of 'Problems in the extreme value analysis'. *Structural Safety* **34** (1), 418–423.

Cunnane, C. 1978 Unbiased plotting positions – A review. *Journal of Hydrology* **37** (3–4), 205–222.

Boughton, W. & Droop, O. 2003 Continuous simulation for design flood estimation – a review. *Environmental Modelling & Software* **18** (4), 309–318.

Gringorten, I. 1963 A plotting rule for extreme probability paper. *Journal of Geophysical Research* **68** (3), 813–814.

Gumbel, E. 1958 *Statistics of Extremes*. Columbia University Press, New York.

Hosking, J. & Wallis, J. 1997 *Regional Frequency Analysis: An Approach Based on L-moments*. Cambridge University Press, New York.

Hyndman, R. & Fan, Y. 1996 Sample quantiles in statistical packages. *American Statistician* **50**, 361–365.

Kjeldsen, T. 2010 Modelling the impact of urbanization on flood frequency relationships in the UK. *Hydrology Research* **51** (5), 391–405.

Kjeldsen, T. R. & Prosdocimi, I. 2015 A bivariate extension of the Hosking and Wallis goodness-of-fit measure for regional distributions. *Water Resources Research* **51** (2), 896–907.

Kjeldsen, T., Jones, D. & Bayliss, A. 2008 *Improving the FEH Statistical Procedures for Flood Frequency Estimation, R&D Report SC050050/TR*. Environment Agency, Bristol.

Makkonen, L. 2008 Bringing closure to the plotting position controversy. *Communications in Statistics – Theory and Methods* **37** (3), 460–467.

Moore, R. J. & Clarke, R. T. 1981 A distribution function approach to rainfall-runoff modelling. *Water Resources Research* **17** (5), 1367–1382.

Packman, J. 1980 *The Effect of Urbanisation on Flood Magnitude and Frequency*. Institute of Hydrology Report No. 63, Institute of Hydrology, Wallingford, UK.

R Core Team 2014 *R: A Language and Environment for Statistical Computing*. R Foundation for Statistical Computing, Vienna, Austria. http://www.R-project.org/.

Robson, A. & Reed, D. 1999 Statistical procedures for flood frequency estimation. *Flood Estimation Handbook*, Vol. 3. Institute of Hydrology, Wallingford, UK.

Shaw, E. 1993 *Hydrology in Practice*, 3rd edn. Stanley Thornes, Cheltenham.

Sooyoung, K., Hongjoon, S., Joo, K. & Heo, J. 2012 Development of plotting position for the general extreme value distribution. *Journal of Hydrology* **475**, 259–269.

Stedinger, J. R., Vogel, R. M. & Foufoula-Georgiou, E. 1993 Frequency analysis of extreme events. In: *Handbook of Hydrology* (D. R. Maidment, ed.). Chapter 18, McGraw-Hill, New York.

Wallingford Hydro Solutions. WINFAP 4, WHS. Available from: https://www.hydrosolutions.co.uk/software/winfap-4/ (accessed 29 April 2019).

Wass, P. 2017 *Darent and Cray Hydrological Assessment*. Environment Agency, Bristol.

Wilks, D. (2011). *Statistical Methods in the Atmospheric Sciences*, 3rd edn. International Geophysical, Volume 100. Elsevier, Oxford.

First received 29 November 2018; accepted in revised form 20 August 2019. Available online 25 September 2019

Revising the BFIHOST catchment descriptor to improve UK flood frequency estimates

Adam Griffin, Andy Young and Lisa Stewart

ABSTRACT

The estimate of the base flow index (BFI) based on the Hydrology of Soil Types (HOST) classification, BFIHOST, provides a measure of catchment responsiveness. BFIHOST is used with other variables to estimate the median annual maximum flood (QMED) in the UK standard Flood Estimation Handbook (FEH) statistical method and is also an explanatory variable in ReFH2, the FEH design hydrograph package. The current estimates of BFIHOST are derived from a restricted linear model, and a number of issues in the catchment dataset have been identified since the original work in 1995. The BFI calculated through base flow separation tends to be underestimated in clay-dominated catchments, and the calculation technique performs poorly in ephemeral catchments or those with missing data. The pragmatic bounding of BFI coefficients for permeable soils overlying aquifer outcrops is also problematic for small catchments. This paper investigates alternative regression methods to improve base flow estimates using the HOST class data for 991 stations (compared to 575 in the original); beta regression was found to give the best performance. Combining multiple rare classes into single classes is also shown to improve performance. The new version of BFIHOST was applied to the QMED equation, showing improved performance.

Key words | base flow, catchment hydrology, regression methods, statistical hydrology

Adam Griffin (corresponding author)
Lisa Stewart
Centre for Ecology & Hydrology,
Benson Lane, Crowmarsh Gifford, Oxfordshire, UK
E-mail: *adagri@ceh.ac.uk*

Andy Young
Wallingford HydroSolutions,
Benson Lane, Crowmarsh Gifford, Oxfordshire, UK

INTRODUCTION

The base flow index (BFI) is a widely applied, broad-scale measure used in the UK for measuring variation in the low-flow regimes of gauged catchments. The index was originally proposed in the 1980s by Gustard *et al.* (1980) and is a simple separation algorithm that disaggregates the daily mean flow record for a catchment into low- and high-frequency components. This is achieved by partitioning the daily mean flows into 5-day blocks and identifying the minimum within each 5-day series. The line connecting these minima represents the storage-driven component of the hydrograph. The BFI is the ratio of the volume of this base flow to the total flow. This is shown for two catchments in Figure 1: a permeable, groundwater-dominated chalk catchment in the south of England and an impermeable catchment from the west of Scotland. The BFI is constrained to lie in the interval [0,1] and ranges from around 0.11 for perennial streams draining impermeable catchments to 0.98 for streams draining the most permeable catchments.

doi: 10.2166/nh.2019.166

The HOST classification and BFIHOST

It is difficult to overstate the importance of soils and underlying geology in influencing the movement of water through the landscape at both the site and catchment scales. The Hydrology of Soil Types (HOST) classification of the UK was developed in the mid-1990s to provide a hydrologically relevant classification of soils and parent geology to aid hydrological studies and analyses within the UK (Boorman *et al.* 1995). The original objective was to replace the five-class Winter Rainfall Acceptance Potential mapping of UK soils, developed in the 1970s following the Soil Survey Field Handbook classification system (Hodgson 1974).

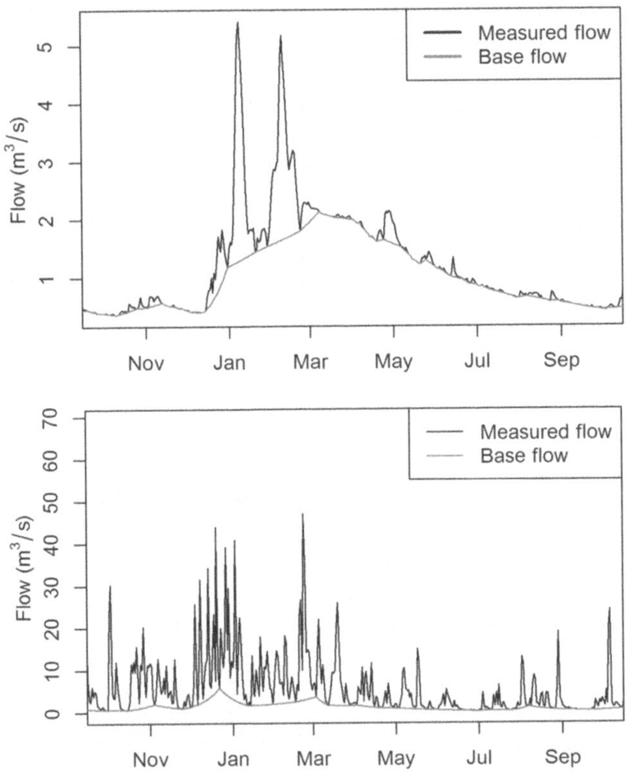

Figure 1 | The base flow separation algorithm of Gustard *et al.* (1992) applied to two catchments for the water year 2013–2014. Top: Pang at Pangbourne (groundwater-dominated). Bottom: Falloch at Glen Falloch (impermeable upland catchment).

The HOST classification is based on a set of conceptual models of the hydrological processes taking place within the soils and, where relevant, the underlying geologies. The classification is based on soil series, and within these settings, soils were differentiated through soil properties and wetness regimes, as indicated by the presence of gleying. This differentiation gave rise to 11 models that were subdivided into 29 classes based on either the geology of the substrate or other properties. The assignment of soil series to these classes was undertaken using a method based on *K*-means cluster analysis. Although HOST is based on soil series, for refinement of the classification, it was applied as a national coverage using the 1:250,000 soil associations (of series) through reconnaissance mapping undertaken by the soil survey organisations across the UK at that time. The conceptual classification of soils was refined by assessing its ability to explain the variation in hydrological characteristics as expressed by the BFI estimated from data for a sample of 575 gauged catchments across the UK. These catchments were selected on the basis of the quality of the hydrometry of the gauged record and the naturalness of the flow regime using the classification scheme of Gustard *et al.* (1992). A linear regression model of the relationship between the BFI and the fractional extents of HOST classes across each catchment in this sample set was then developed, and the resulting BFIHOST catchment descriptor (here referred to as $BFIHOST_{1995}$) was developed by applying this model across the UK to a 1 km grid of the fractional extents of each of the 29 HOST classes and taking an area-weighted average over each catchment.

Figure 2 presents the HOST classification scheme, together with the model coefficients for BFIHOST and the percentage of the UK land surface that each class represents. In general, catchment permeability grades from left to right and top to bottom, and on the whole, this pattern is reflected in the BFIHOST model coefficients, with some notable exceptions. For example, the drained and eroded peats of classes 11 and 28 had higher BFIHOST than surrounding classes, not following this trend. To ensure model coefficients lay in an acceptable range, the regression was also bounded to prevent the coefficients taking values greater than 1 or less than 0.17. In the final model, additional constraints were placed on the sandstone and sand-dominated HOST classes 3 and 5, such that the coefficient value for both classes was constrained; class 3 restricted to having a value above 0.9 and class 5 capped to having a value below 0.9.

Notwithstanding these criticisms, BFIHOST has been a well-used catchment descriptor. It is a key descriptor in the Flood Estimation Handbook (FEH) catchment descriptor equation for estimating the index variable QMED, the median of the annual maximum flood series, in ungauged catchments (Kjeldsen *et al.* 2008) and informs three out of the four model parameters in the design package for the ReFH2 event-based rainfall-runoff model (Wallingford Hydrosolutions 2016a). It has also proved to be a key catchment descriptor for explaining the variation in parameter values for generalised rainfall-runoff models for estimating both daily mean flows (Young 2006; Young *et al.* 2006) and extreme flood events (Calver *et al.* 1999).

The BFIHOST catchment descriptor has also found wide application in environmental management, for example, in the development of environmental standards

Substrate hydrogeology	MINERAL SOILS						PEAT SOILS	
	Groundwater or aquifer	No impermeable or gleyed layer within 100cm	Impermeable layer within 100cm or gleyed layer at 40-100cm		Gleyed layer within 40cm		All	
Weakly consolidated, microporous, by-pass flow uncommon (Chalk)	Normally present and at >2m	[1] 1.000 *(4.31)*	[13] 1.000 *(0.87)*		[14] 0.380 *(0.66)*		[15] 0.380 *(9.93)*	
Weakly consolidated, microporous, by-pass flow uncommon (Limestone)		[2] 1.000 *(2.12)*						
Weakly consolidated, macroporous, by-pass flow uncommon		[3] 0.900 *(1.58)*						
Strongly consolidated, non or slightly porous, by-pass flow common.		[4] 0.791 *(3.33)*						
Unconsolidated, macroporous, by-pass flow very uncommon.		[5] 0.900 *(5.07)*						
Unconsolidated, microporous, by-pass flow very common.		[6] 0.645 *(2.61)*						
Unconsolidated, macroporous, by-pass flow very uncommon.	Normally present and at ≤ 2m	[7] 0.792 *(1.01)*			IAC' <12.5	IAC' ≥12.5	Drained	Undrained
Unconsolidated, microporous, by-pass flow common.		[8] 0.560 *(1.62)*			[9] 0.734 *(3.68)*	[10] 0.520 *(2.21)*	[11] 0.927 *(0.55)*	[12] 0.170 *(2.94)*
Slowly permeable	No significant groundwater or aquifer	[16] 0.778 *(0.43)*	IAC' >7.5	IAC' ≤7.5	[24] 0.312 *(13.85)*		[26] 0.244 *(2.49)*	
			[18] 0.518 *(5.40)*	[21] 0.340 *(4.02)*				
Impermeable (hard)		[17] 0.609 *(9.28)*	[19] 0.469 *(2.16)*	[22] 0.315 *(1.10)*			[27] 0.259 *(0.83)*	
Impermeable (soft)			[20] 0.524 *(0.69)*	[23] 0.218 *(1.31)*	[25] 0.170 *(3.64)*			
Eroded Peat							[28] 0.581 *(0.58)*	
Raw Peat							[29] 0.226 *(5.73)*	

Figure 2 | HOST classification (small upper numbers) with BFIHOST coefficients listed (percentage land cover in brackets) based on Boorman *et al.* (1995).

for abstraction as part of the implementation of the Water Framework Directive in the UK (SNIFFER 2006). Another example is its use in understanding the role of hydrology in explaining the interactions between dissolved organic carbon and nitrogen in rivers (Heppell *et al.* 2017).

Operational use of research outputs often identifies limitations, and the BFIHOST model is no exception. In impermeable catchments (low BFI) such as those found in upland peat-dominated catchments, $BFIHOST_{1995}$ tends to overestimate at low gauged BFI values (Figure 3). This can be seen in Figure 2 in the high coefficient for class 11 compared with the overall trend. On the other hand, HOST classes 23 and 25, which represent clay-based soils, show much lower values of BFIHOST than the overall trend would suggest. In such impermeable catchments, those with a BFI less than 0.2 are often associated with the ephemerality and zero flow in summer months.

Another problem associated with some HOST classes is linked to their relative rarity within the dataset. For example, HOST class 11 is only observed in 0.55% of the original 1995 dataset, and was also seen to be frequently

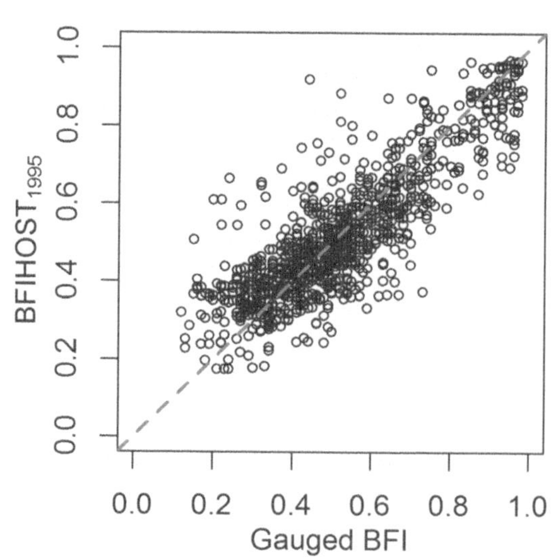

Figure 3 | Comparison of gauged BFI estimates with $BFIHOST_{1995}$ (Boorman *et al.* 1995) using the present dataset of 991 catchments.

co-located with permeable soils, and so is likely poorly modelled by $BFIHOST_{1995}$ due to having some possible correlation with other soil types. It is hoped that using a larger more varied dataset may alleviate this.

Finally, the need to artificially change model coefficients (in the cases of classes 1, 2, 3, 5 and 13) suggests that the model was not chosen carefully enough in the outset. An important part of model development is the choice of the model form, be that linear regression, a more generalised additive model or even a more physically based mathematical hydrological model. The parameter estimates in permeable soil classes represent a theoretical upper limit where the parameter values take a bounded value of 1. In practice, BFI estimates from gauged records may approach this theoretical limit, but even the flow regimes of the most permeable catchments exhibit some direct response to rainfall, and thus a value of 1 is not attained.

In this paper, the aim is to tackle these issues and produce a more robust model of BFI. To achieve this, a more physically appropriate model will be investigated to avoid the need for capping coefficients. Secondly, methods will be investigated to improve the base flow estimates of the rarer HOST classes, by combining single HOST classes into groups. Thirdly, models focusing on the low BFI catchments will be explored to improve the estimates in the peat and clay catchments. This will be then demonstrated through an application to the estimation of QMED through the FEH catchment descriptor equation.

DATA

Station selection

The catchment dataset used in the 1995 development of the HOST classification was constrained to using catchments with flow records that were believed to be of good hydrometric quality and relatively free from artificial influence (Gustard *et al.* 1992) and applied to all gauged flow records held by the National River Flow Archive (NRFA). However, information on water use and return were fragmented in England and Wales, and in Scotland and Northern Ireland, there was no requirement to regulate abstraction until 2006. A necessarily conservative approach was taken for the selection of study catchments. Due to differences in the mapping and assignment of HOST classes within Northern Ireland compared with England, Scotland and Wales, catchments in Northern Ireland were not included. Instead, we feel that future work should be undertaken to properly develop a Northern Ireland-specific BFIHOST model.

A focus of the current study was to develop a much larger sample set of appropriate gauged records to inform the research. The NRFA holds flow records for the UK gauged catchments judged by the measuring authorities as a suitable hydrometric quality for public release. On the basis that BFI is relatively insensitive to hydrometric quality (pp. 24, Gustard *et al.* 1992), all stations were considered as being of suitable hydrometric quality. The records for 1,223 catchments with a minimum of 10 years of daily mean flow record were reviewed and 991 catchments selected according to the following criteria:

- the absence of significant upstream impounding reservoirs;
- less than 2% missing days in the record;
- generally, perennial stream flow, determined on the basis that the flow that is equalled or exceeded for 95% of the time, Q95, is greater than 0.

Lakes were not chosen as criteria for exclusion due to the very high percentage of catchments with some form of surface water storage. Instead, a HOST class (class 30) describing the proportion of lake coverage is included; this is explained in more detail below. Previous research in Scotland (Gustard *et al.* 1987) has shown that the estimation of BFI from gauged records is relatively insensitive to sample error and that, provided that extremely dry years are avoided, the error in the BFI calculated from 1 year of the record is typically approximately 5%. However, BFI values are sensitive to missing data within the period of calculation and particularly within periods of low flow; a limit on missing data of 2% was proposed by Gustard *et al.* (1992), and this motivated the use of this same limit in the present study.

Abstractions and discharges in the UK are, in the main, regulated to maintain the magnitude of influence to below a fraction of the lower flows observed within a catchment. When compared with the full range of variation in the daily mean flow records for a catchment (typically three to four orders of magnitude), these influences are relatively invariant. Errors at very high flows associated with out-of-bank flow are not as relevant as those which occur in a small fraction of the higher-frequency components of the flow record. In contrast, impounding reservoirs have a significant impact on the whole flow regime.

It was noted that some of the 991 gauged flow records showed non-negligible artificial influence compared to Q95. In particular, 332 catchments were seen to have net influence estimated to be greater than 20% of Q95. Fortunately, filtering out ephemeral catchments also screens out catchments with a very high net abstractive component. Sensitivity to this abstraction was investigated. The basic linear model was fitted with and without those 332 catchments with noticeable abstraction, and 10-fold cross-validation was performed in both cases (James *et al.* 2013). It was seen that the basic model performed similarly with and without the extra catchments, both in terms of model residual and cross-validation prediction errors. Therefore, the rest of this paper will make use of all 991 flow records, as it is felt that a dataset that covers a broader range of catchments will lead to a model which is more representative of base flow across the whole UK. This does come at the slight cost of increased observation uncertainty, but as mentioned above, Gustard *et al.* (1992) show that the BFI is robust to this sort of data issue.

Figure 4 shows the locations of the catchments within the UK. The mean record length of stations within the sample is 42.3 years. Catchment areas range from 0.9 km^2 up to 9,940 km^2, with a mean area of 343 km^2.

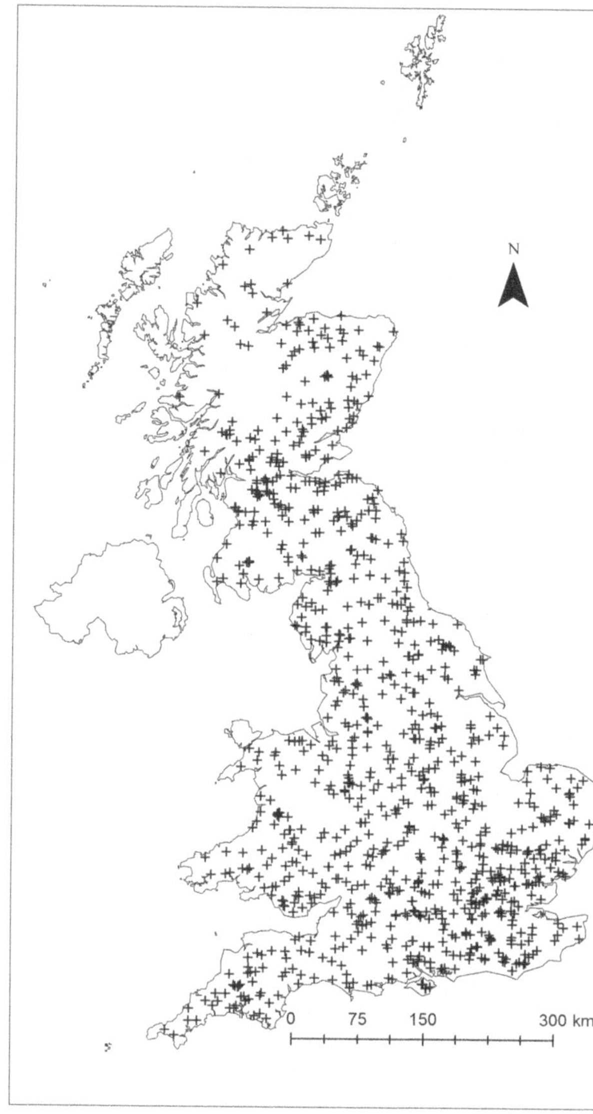

Figure 4 | Locations of 991 catchments selected for new models of BFIHOST.

METHODS

As mentioned above, $BFIHOST_{1995}$ was a linear regression model with bounds placed on the minimum (0.17) and maximum (1.0) values BFIHOST could take for any specific HOST class (Boorman *et al.* 1995). In the first stage of the current study, the linear model was recalibrated using the new dataset and then directly compared with $BFIHOST_{1995}$. Simple linear regression models are easy to interpret, but have problems with extrapolation; the application of them here possibly leads to estimated values of BFIHOST greater than 1 or less than 0. Catchment estimates are given by $BFIHOST = \sum_{i=1}^{29} \alpha_i h_i$ where α_i is the BFI coefficient for HOST class i and h_i is the proportion of that catchment which is classified as belonging to HOST class i. This assumes that the error (BFI_{gauged} − BFIHOST) is normally distributed, which may suggest that the true value could lie outside the permitted range even if the fitted model estimate lies within the range.

To more directly compare to $BFIHOST_{1995}$, the recalibrated linear model was also adjusted to cap coefficients to lie in the range of $0 < \alpha < 1$. Within-range capping was not considered in this work. Linear models were implemented using the core *stats* R package (R Core Team 2016).

Beta regression

As an alternative to the simple linear regression, a beta regression model was investigated. Rather than assuming that the residual follows a normal distribution as in linear regression, beta regression assumes that each observation comes from a beta distribution with mean fitted using a logit link function (Ferrari & Cribari-Neto 2004). Here, $\mathrm{logit}(x) = \log(x/(1-x))$, a function that turns the unit interval (0, 1) into the whole range from $-\infty$ to ∞; inverting the transformation naturally forces the resulting estimates of BFIHOST to lie strictly between 0 and 1. More specifically, the gauged values of the BFI at site i are assumed to be distributed according to a beta distribution with mean μ_i and precision ϕ common to all sites. Here, μ_i is given by the following equation: $\mathrm{logit}(\mu_i) = \sum_i \beta_i h_i$, where β_i are fitted model coefficients and h_i are as defined previously. The beta distribution is completely defined between 0 and 1 and has the probability density function given by:

$$f(x) = \frac{\Gamma(\phi)}{\Gamma(\phi\mu)\Gamma(\phi(1-\mu))} x^{\mu\phi-1}(1-x)^{(1-\mu)\phi-1}, \qquad 0 < x < 1$$

where $\Gamma(x)$ is the gamma function. BFIHOST is then estimated by using:

$$\mathrm{BFIHOST_{BETA}} = \mathrm{logit}^{-1}\left(\sum_i \beta_i h_i\right)$$

with β_i and h_i as defined earlier. The beta regression was implemented using the *betareg* R package (Cribari-Neto & Zeileis 2010). In order to better compare the coefficients directly, a set of 'linear-equivalent' values will be reported, corresponding to the value b_i of $\mathrm{BFIHOST_{BETA}}$ in a catchment consisting solely of a single HOST class i, which can be calculated as $b_i = \mathrm{logit}^{-1}(\beta_i)$.

One benefit of using a beta regression is that it is designed for estimating proportions, and so the estimate plus the associated uncertainty (the error term) still lies within the interval of appropriate BFI values.

Regrouping variables

The 1995 BFIHOST model was a classification tool to aid the development of HOST, whereas the objective of this study is to refine the estimation of BFI to give reliable estimates across all catchment scales and with HOST class representations ranging from a single dominant class to large catchment draining soils corresponding to many different HOST classes.

A 30-class model including a HOST class representing surface water extents (HOST class 30) is considered; nominally, one would expect the BFI of a waterbody to be 1, as none of the water would flow over the top of a waterbody, but rather through it. The effect of surface water, in the form of lakes and reservoirs (for example), is already well established in flood frequency estimation using catchment descriptor equations, where the extent to which such water bodies attenuate flow is significant in the QMED equation (Kjeldsen *et al.* 2008). This 30-class model was compared to the 29-class model under the original 1995 coefficients and under the recalibrated linear model.

Typically, the BFI for catchments which are dominated by rarer HOST classes is over- or underestimated in $\mathrm{BFIHOST_{1995}}$, since there are many fewer catchments in which those HOST classes are observed. For example, HOST class 11 only makes up 0.55% of the land cover in England, Wales and Scotland, and makes up less than 11.2% of any one catchment in the present dataset.

To analyse this, a series of models were tested where subsets of the 30 classes were combined replacing, for example, HOST classes 16 and 17 with a single (16 + 17) class (so $h_{(16+17)} = h_{16} + h_{17}$). Various combinations of HOST class groupings were investigated, as outlined in Table 2. Each of these groupings was chosen to combine similar soil classes (in terms of base flow) together to address the poor representation of one or more classes within the group. This was then followed by considering combinations of the above single groups to develop the final model.

Weighted models

The most successful models were tested using weighted regression models, where certain data points were given greater weighting in the regression method (either linear or beta). Catchments with a gauged BFI value of less than 0.4 were seen to perform particularly badly in the original 1995 model, and so a more focused approach to characterising the BFI in these catchments was felt to be important for use in prediction across the UK.

To develop the weighted models, four methods were considered. The catchments with the BFI less than 0.4 were weighted twice as much, three times as much, and six times as much as those with a higher BFI. Also, a weighting inversely proportional to BFI was tried.

RESULTS AND DISCUSSION

The results in this section show model performance in terms of R^2 (a measure of the variance of the data explained by the model), root mean-squared error (RMSE, a measure of average accuracy of the model), fractional bias, and Akaike Information Criterion (AIC, Akaike 1974), which is given by AIC $= 2k - 2\log(L)$, where k is the number of fitted parameters (the number of classes/groups selected) and L is the maximum value of the likelihood function of the fitted model given the dataset; note that lower or more negative values show better models. One should note that the R^2 values are designed to assume normally distributed residuals about the model fitted values. This means that the values of R^2 associated with the beta regression may incorrectly claim under-performance of the model. Therefore, RMSE and AIC are more appropriate for comparing linear with beta regression models. Within the type, R^2 can be more appropriately used as a comparison.

30-class model

First, an investigation was performed to assess the original model and the benefits of including HOST class 30 (surface water). Table 1 suggests that the old model much better explains the data than previously thought, with even further improvements in terms of R^2 and RMSE. The AIC of the original model cannot be determined due to the lack of access to the original 1995 dataset; the RMSE is as reported in Boorman *et al.* (1995).

Table 1 | Comparison of various linear models

Model	R^2	RMSE	AIC	BIAS
1995 model (as reported)	0.79	0.089	N/A	N/A
1995 model (new dataset)	0.97*	0.101	N/A	+0.050
Recalibrated linear 29-class	0.97	0.096	−1,767	−0.052
Linear 30-class	0.97	0.095	−1,797	−0.049
Capped 30-class	0.97	0.095	−1,776	+0.037

AIC and bias not documented for the original 1995 model. A value of R^2 for the 1995 model under the new data (denoted by *) is only illustrative, not a true value.

However, one should note that this is not an ideal use of R^2 since the data used to calculate the model here was not the data used to calibrate the model in Boorman *et al.* (1995), and so the resultant value of R^2 should be treated with caution. The recalibrated 29-class model is a naïve linear regression and, therefore, some of the coefficients extend beyond 0 and 1; HOST class 1 has a value of 1.02, class 11 has a value of 1.15, and class 13 has a value of 1.19. Including the HOST class 30 (surface water extent) does not improve this, with HOST class 1 taking a coefficient of 1.02, class 11 having a value of 1.17, class 12 having a value of −0.03, and class 27 taking a value of −0.13. This could be a significant problem for estimation in small catchments which predominantly consist of these HOST classes.

Reapplying the capping procedure of Boorman *et al.* (1995) on the new 29- and 30-class linear models only served to reduce the efficacy of the model (Table 1). Bias may have decreased as an artefact of restricting the model to the 'correct' range. In this case, the model statistics were recomputed, assuming that the error term had the same distribution in the capped and uncapped 30-class linear model.

Beta regression as an alternative

Table 2 shows the initial results of using the beta regression versus the linear model; the bias is slightly improved, and the values of R^2 and RMSE are very similar. Recall that since linear regression coefficients and beta regression coefficients cannot be directly compared, Table 2 presents 'linear-equivalent' values. These correspond to the value b_i of $\text{BFIHOST}_{\text{BETA}}$ in a catchment consisting solely of a single HOST class i, which can be calculated as $b_i = \text{logit}^{-1}(\beta_i)$, where β_i is the beta regression coefficient for HOST class i.

Table 4 shows that the beta regression model keeps all the 'linear-equivalent' coefficients between 0 and 1 without further modification so that no catchment can have an estimated value of BFIHOST outside this range, which is desirable. Here, the value of R^2 is computed as for linear regression. In the *betareg* package, a 'pseudo-R^2' specific to beta regression is also

Table 2 | Comparison of beta regression models under different groupings of HOST classes

Model	R^2	RMSE	AIC	Bias
Linear 30-class model	0.970	0.095	−1,797	−0.049
Beta 30-class model	0.970	0.098	−1,819	+0.059
(7 + 8)	0.970	0.095	−1,821	+0.052
(9 + 10)	0.970	0.095	−1,822	+0.052
(11 + 12 + 15)	0.969	0.096	−1,800	+0.053
(13 + 14)	0.970	0.095	−1,816	+0.052
(16 + 18)	0.970	0.095	−1,812	+0.052
(20 + 23)	0.970	0.095	−1,822	+0.052
(26 + 27)	0.970	0.095	−1,815	+0.052
(28 + 29)	0.970	0.095	−1,819	+0.052
(7 + 8 + 9 + 10)	0.970	0.095	−1,825	+0.052
(7 + 8 + 9 + 10, 26 + 27)	0.970	0.095	−1,820	+0.052
(7 + 8 + 9 + 10, 11 + 12 + 15, 26 + 27)	0.969	0.096	−1,800	+0.054
(7 + 8 + 9 + 10, 11 + 12 + 15, 16 + 18, 20 + 23, 26 + 27)	0.968	0.096	−1,794	+0.054

Bracketed groupings replace the individual constituent classes, and all other unlisted classes are kept individually.

presented (Ferrari & Cribari-Neto 2004). To more appropriately compare between the linear and beta regression models, the linear regression formulation of R^2 is used.

Combined HOST classes

Table 2 shows the difference in including various combinations of the groupings outlined above. Here, one notes that certain models perform slightly better (the (9 + 10) model is slightly better on all statistics), but upon examining the coefficients and the linear model-equivalent values, it can be observed that they produce physically unrealistic values close to 0 or 1: HOST class 12 has a 'linear-equivalent' coefficient of 0.074, and class 27 has a value of 0.068, much lower than reasonable. As a balance between physically reasonable and statistically 'powerful' models must be reached, this led to the selection of a model using the grouped classes (7 + 8 + 9 + 10, 11 + 12 + 15, 16 + 18, 20 + 23, 26 + 27), with all other classes kept in isolation. In the rest of this paper, only the 30-class and (7 + 8 + 9 + 10, 11 + 12 + 15, 16 + 18, 20 + 23, 26 + 27) models will be considered; the latter will be referred to as the '22-class' model.

Weighted models

Finally, the selected 22-class model and the 30-class model under linear and beta regression were investigated using the weighting scheme outlined in the Methods section. Small relative weightings were tried but showed little difference compared with the unweighted model (not shown), and very high relative weightings instead produced a poor fit for high BFI stations (Figure 5). Finally, a weighting inversely proportional to the BFI was also tried, but this also gave far too much weight to the low BFI catchments and led to a poorly fitting model (not shown). Hence, the four candidate models were compared with similar models weighting catchments with a gauged BFI between 0.17 and 0.4 three times as heavily as other catchments. Although R^2 and RMSE do not greatly change, continued improvement can be observed in Table 3 in terms of AIC. The final weighted 22-class beta regression model is denoted as $BFIHOST_{2019}$

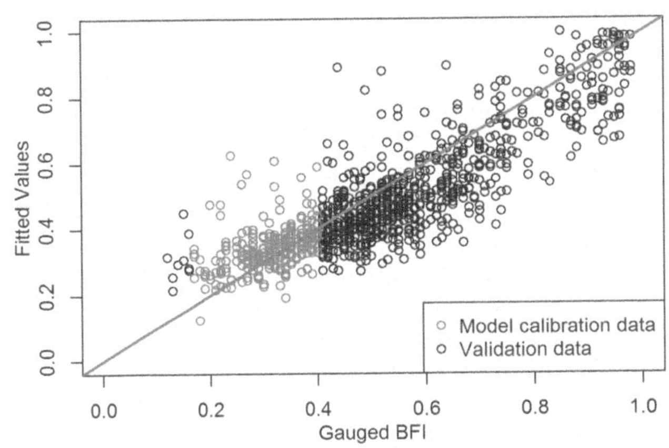

Figure 5 | Plots showing the accuracy of fit underweighting for linear 30-class model fitted with catchments with a gauged BFI between 0.17 and 0.4 weighted six times as heavily as those outside the range.

Table 3 | Comparison of various weighted regression models under linear and beta formulations and under 22- and 30-class configurations

Model	R^2	RMSE	AIC	Bias
Weighted linear 30-class model	0.962	0.098	−1,687	+0.039
Weighted beta 30-class	0.968	0.0975	−2,923	+0.040
Weighted linear 22-class	0.960	0.100	−1,657	−0.001
Weighted beta 22-class	0.967	0.097	−1,798	+0.006

Table 4 | HOST coefficients for linear models and 'linear-equivalent' values for beta regression models

Host	$BFIHOST_{1995}$	Linear 30-class	Beta 30-class	Linear 22-class + weighted	Beta 22-class + weighted	Host	$BFIHOST_{1995}$	Linear 30-class	Beta 30-class	Linear 22-class + weighted	Beta 22-class + weighted
1	1.000	1.031	0.956	1.035	0.949	16	0.778	0.617	0.559	0.491	0.492
2	1.000	1.007	0.927	0.938	0.881	17	0.609	0.564	0.564	0.514	0.514
3	0.900	0.690	0.671	0.709	0.704	18	0.518	0.485	0.480	0.491	0.492
4	0.791	0.819	0.793	0.773	0.756	19	0.469	0.456	0.467	0.214	0.234
5	0.900	0.974	0.888	0.982	0.878	20	0.524	0.390	0.233	0.319	0.302
6	0.645	0.463	0.486	0.367	0.402	21	0.340	0.354	0.333	0.328	0.323
7	0.792	0.830	0.756	0.718	0.682	22	0.315	0.398	0.397	0.369	0.374
8	0.560	0.756	0.825			23	0.218	0.137	0.119	0.319	0.302
9	0.734	0.884	0.705			24	0.312	0.317	0.326	0.325	0.333
10	0.520	0.634	0.561			25	0.170	0.247	0.258	0.190	0.209
11	0.927	1.608	0.984	0.266	0.271	26	0.244	0.272	0.287	0.222	0.249
12	0.170	0.010	0.090			27	0.259	−0.184	0.055		
13	1.000	1.124	0.933	1.224	0.955	28	0.582	0.455	0.474	0.424	0.447
14	0.380	0.660	0.671	0.664	0.702	29	0.226	0.253	0.257	0.246	0.254
15	0.380	0.296	0.304	0.266	0.271	30	N/A	2.263	1.000	1.583	0.991

Shaded regions correspond to combined classes.

To summarise the findings, Table 4 and Figure 6 show the coefficients from the linear regression models and the 'linear-equivalent' values from the beta regression models, respectively. For the less frequently observed HOST classes, a more reasonable value of BFIHOST is obtained when combined with a more abundant HOST class; classes 11 and 12 are a clear example of this (with a land cover of 0.55% and 2.94%, respectively). Figure 7 shows a residual plot as described in Chien (2011), which performs in a similar way to a standard residual plot as used for linear regression models. There is no obvious trend or pattern to the residuals as a function of logit($BFIHOST_{2019}$), which suggests that the model type is appropriate for this data.

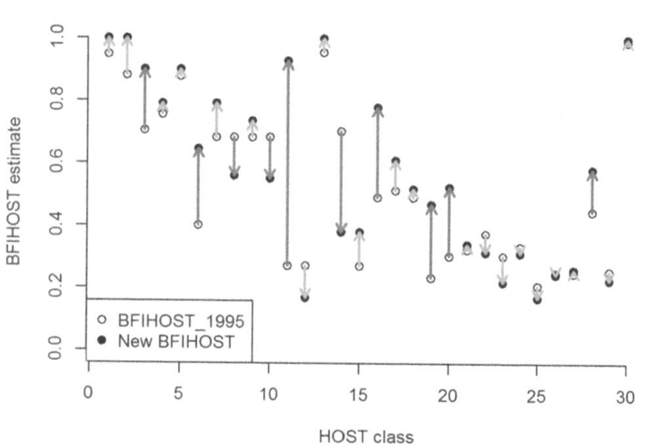

Figure 6 | Comparison of BFIHOST estimates of single-class catchments, with strong differences highlighted with upward arrows (new BFIHOST much larger) and downward arrows (new BFIHOST much smaller).

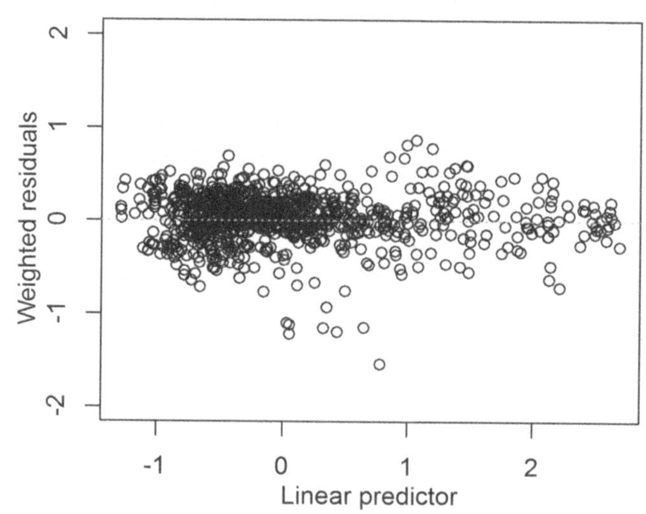

Figure 7 | Beta regression residual plot showing scaled residuals against estimates (linear predictors).

Example application to flood estimation

As a simple application of $BFIHOST_{2019}$, a standard use of BFIHOST will be investigated: the estimation of QMED. To do this, a subset of the above investigated dataset has been used: 605 stations which are rural and determined by the NRFA to be suitable for QMED estimation. Here, $BFIHOST_{1995}$ and the new $BFIHOST_{2019}$ were both used in the QMED catchment descriptor equation without recalibration:

$$QMED = 8.3062\ AREA^{0.8510}\ 0.1536^{\frac{1000}{SAAR}} FARL^{3.4451}\ 0.046^{(BFIHOST^2)}$$

where AREA is the area of the catchment in km^2, SAAR is the standardised annual average rainfall in mm (based on data from 1961 to 1990), FARL is a coefficient describing attenuation due to lakes and reservoirs, and QMED is measured in $m^3\ s^{-1}$ (Kjeldsen *et al.* 2008). Figure 8 shows the values estimated for QMED under the two methods of deriving BFIHOST. Here, it can be seen that the value for QMED under the new $BFIHOST_{2019}$ performs slightly better, particularly for catchments with smaller values of QMED (comprising smaller catchments and more permeable catchments), where the QMED model typically performs less well (Vesuviano *et al.* 2016).

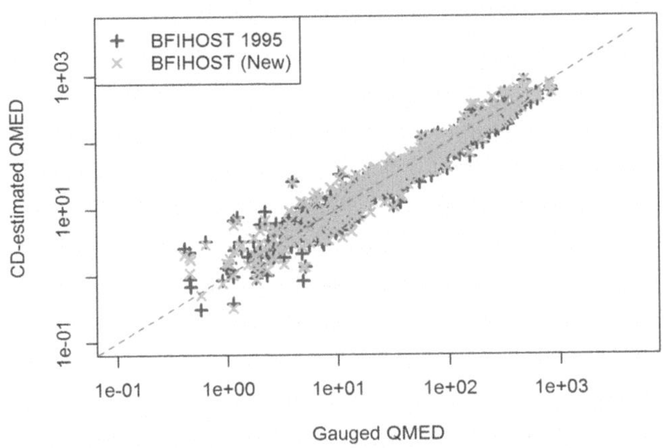

Figure 8 | Comparison of fit of Catchment Descriptor (CD) QMED equation under the old BFIHOST from Boorman et al. (1995), and the new BFIHOST model.

Table 5 shows that the estimation of QMED and BFIHOST in impermeable catchments (BFI < 0.4) is improved (in terms of factorial standard error) by the new model at all catchment sizes, but performance is very similar across sizes of catchments. Hence, $BFIHOST_{2019}$ can address the concerns about parameter estimates for specific scarce HOST classes without a loss of estimation performance when used in the context of generally larger catchments. It should be noted that the equation was calibrated using the $BFIHOST_{1995}$ estimates and thus further model improvement may be gained by recalibrating the equation using $BFIHOST_{2019}$.

CONCLUSIONS

This paper has investigated the potential for an updated method of estimating base flow at ungauged locations, improving on $BFIHOST_{1995}$ by applying a beta regression model instead of the original capped linear regression model, in addition to using a new larger dataset of over 900 gauged catchments.

Table 5 | Description of factorial standard error of QMED and BFI under $BFIHOST_{1995}$ and $BFIHOST_{2019}$ for small/large and permeable/impermeable catchments

Area	BFI	QMED fse $BFIHOST_{1995}$	QMED fse $BFIHOST_{2019}$	BFI fse $BFIHOST_{1995}$	BFI fse $BFIHOST_{2019}$
All	All	1.535549	1.531486	1.102568	1.103955
<40 km^2	All	1.743654	1.755493	1.123471	1.128958
>40 km^2	All	1.499231	1.492109	1.098892	1.099484
All	<0.4	1.479287	1.450971	1.103977	1.083578
All	>0.4	1.555649	1.559788	1.102043	1.11066
<40 km^2	<0.4	1.692531	1.658577	1.121898	1.102398
<40 km^2	>0.4	1.792398	1.846157	1.12499	1.151069
>40 km^2	<0.4	1.397002	1.369949	1.097331	1.076356
>40 km^2	>0.4	1.52839	1.526364	1.099365	1.105668

Choosing a beta regression allowed a model to be fitted which naturally gives estimates strictly between 0 and 1, avoiding hydrologically unrealistic estimates for the BFI. Some HOST classes do not occur in any great quantities at any location or are highly concentrated in an extremely small number of locations. In locations where these are present, base flow is often poorly estimated by $BFIHOST_{1995}$.

This issue was still present in the original beta regression model, giving incredibly high/low values for these classes, due to insufficient information to fit accurately. To this end, HOST classes were combined, grouping rare classes with HOST classes that are more abundant and have similar physical and hydrological properties (Table 1). This led to a 22-class model which gave a good fit without a need to impose artificial constraints on the model parameterisation.

To demonstrate its applicability and validity, this new $BFIHOST_{2019}$ was used in the existing QMED catchment descriptor equation. The resultant estimates are an improvement over the use of the equation with the original BFIHOST estimates, despite the equation having been fitted using those original BFIHOST estimates. This is in addition to the core objective of resolving unrealistic estimation in small catchments dominated by single HOST class values that were poorly represented in the dataset used in the original model development. To extend this work, it would be fruitful to recalibrate the QMED equation using the generalised linear model developed in Kjeldsen *et al.* (2008) and also to recalibrate the parameters for the ReFH2 model (Wallingford Hydrosolutions 2016b), namely C_{max}, BL and BR.

ACKNOWLEDGEMENTS

The authors thank the British Hydrological Society for the opportunity to publish this work. The authors also thank the editors and reviewers for their helpful comments.

REFERENCES

Akaike, H. 1974 A new look at the statistical model identification. *IEEE Transactions on Automatic Control.* doi:10.1109/TAC.1974.1100705.

Boorman, D., Hollis, J. M. & Lilly, A. 1995 *Hydrology of Soil Types: A Hydrologically-Based Classification of the Soils of United Kingdom.* Institute of Hydrology, Wallingford. doi:10.1029/98GL02804.

Calver, A., Lamb, R. & Morris, S. E. 1999 Flood frequency estimation using continuous rainfall-runoff modelling. *Proceedings of the Institute of Civil Engineers – Water and Maritime Engineering* **136**, 225–234. doi:10.1016/S0079-1946(96)00010-9.

Chien, L. C. 2011 Diagnostic plots in beta-regression models. *Journal of Applied Statistics* **38** (8), 1607–1622. doi:10.1080/02664763.2010.515677.

Cribari-Neto, F. & Zeileis, A. 2010 Beta regression in R. *Journal of Statistical Software* **34** (2), 1–24. Retrieved from: http://www.jstatsoft.org/v34/i02.

Ferrari, S. & Cribari-Neto, F. 2004 Beta regression for modelling rates and proportions. *Journal of Applied Statistics* **31** (7), 799–815. doi:10.1080/0266476042000214501.

Gustard, A., Bullock, A. & Dixon, J. M. 1980 *Low Flow Estimation in the United Kingdom (Vol. IH Report).* Institute of Hydrology, Wallingford.

Gustard, A., Bullock, A. & Dixon, J. M. 1992 *Low Flow Estimation in the United Kingdom.* Institute of Hydrology, Wallingford.

Gustard, A., Marshall, D. C. W. & Sutcliffe, M. F. 1987 *Low Flow Estimation in Scotland.* Institute of Hydrology, Wallingford.

Heppell, C. M., Binley, A., Trimmer, M., Darch, T., Jones, A., Malone, E., Collins, A. L., Johnes, P. J., Freer, J. E. & Lloyd, C. E. M. 2017 Hydrological controls on DOC nitrate resource stoichiometry in a lowland, agricultural catchment, southern UK. *Hydrology and Earth System Sciences* **21** (9), 4785–4802. doi:10.5194/hess-21-4785-2017.

Hodgson, J. M. 1974 *Soil Survey Field Handbook: Technical Monograph No. 5.* Rothamsted Experimental Station, Harpenden, UK.

James, G., Witten, D., Hastie, T. & Tibshirani, R. 2013 Classification. *An Introduction to Statistical Learning with Applications in R with Applications in R.* Springer-Verlag, New York, USA. doi:10.1007/978-1-4614-7138-7.

Kjeldsen, T. R., Jones, D. A. & Bayliss, A. C. 2008 *Improving the FEH Statistical Procedures for Flood Frequency Estimation.* Joint Defra/Environment Agency Flood and Coastal Erosion Risk Management R&D Programme, Science Report SC050050, London, UK.

R Core Team 2016 *R: A Language and Environment for Statistical Computing.* Vienna. Retrieved from: https://www.r-project.org/.

SNIFFER 2006 *Development of Environmental Standards (Water Resources). WFD48 Stage 3: Environmental Standards.* Edinburgh. Retrieved from: https://www.sniffer.org.uk/wfd48-stage-3-report-pdf.

Vesuviano, G., Stewart, L., Haxton, T., Young, A., Hunt, T., Spencer, P. & Whitling, M. 2016 Reducing uncertainty in small-catchment flood peak estimation. *E3S Web of Conferences* **7**, 01008. doi:10.1051/e3sconf/20160701008.

Wallingford Hydrosolutions 2016a *The Revitalised Flood Hydrograph Model ReFH2: Technical Guidance.* Wallingford Hydrosolutions Ltd, Wallingford, UK.

Wallingford Hydrosolutions 2016b *WINFAP 4 QMED Linking Equation*. Wallingford Hydrosolutions Ltd, Wallingford, UK.

Young, A. R. 2006 Stream flow simulation within UK ungauged catchments using a daily rainfall-runoff model. *Journal of Hydrology*. doi:10.1016/j.jhydrol.2005.07.017.

Young, A. R., Keller, V. & Griffiths, J. 2006 Predicting low flows in ungauged basins: a hydrological response unit approach to continuous simulation. In: *Climate Variability and Change – Hydrological Impacts*. (S. Demuth, A. Gustard, E. Planos, F. Scatena & E. Servat, eds). International Association of Hydrological Sciences Press, Wallingford, UK, pp. 134–138. doi:10.1063/1.478943.

First received 14 December 2018; accepted in revised form 14 May 2019. Available online 31 May 2019

Fully distributed hydrological modelling for catchment-wide hydrological data verification

D. Ocio, T. Beskeen and K. Smart

ABSTRACT

Hydrological data scarcity and uncertainty is a fundamental challenge in hydrology, particularly in places with weak or declining investment in hydrometric networks. It is well established that fully distributed hydrological models can provide robust estimation of flows at ungauged locations, through local calibration and regionalisation using spatial datasets of physical properties. Even in situations where data are abundant, the existence of inconsistent information is not uncommon. The measurement, estimation or interpolation of rainfall, potential evapotranspiration and flow as well as the difficulty in monitoring artificial influences are all sources of potential inconsistency. Less studied but as important, distributed hydrological models, given their capability of capturing both the temporal and spatial dimensions of the water balance and runoff generation, are suitable tools to identify potential deficiencies in, and reliability of, input data. Three heavily modified catchments in the East of England such as the Ely Ouse, the Witham, and the Black Sluice have been considered, all of which have issues of data scarcity and uncertainty. This paper demonstrates not only the benefits of fully distributed modelling in addressing data availability issues but also in its use as a catchment-wide data validation tool that serves to maximise the potential of limited data and contributes to improved basin representation.

Key words | data uncertainty, data validation, distributed hydrological modelling, ungauged catchments, water resource assessment

D. Ocio
T. Beskeen (corresponding author)
Mott MacDonald,
22 Station Road, Cambridge CB1 2JD,
UK
E-mail: *tom.beskeen@mottmac.com*

K. Smart
Anglian Water Services,
Milton House, Cowley Road, Cambridge CB4 0AP,
UK

INTRODUCTION

The development of hydrological models for the determination of water resource availability is dependent on suitable hydrological data records. In many countries, particularly developing ones, hydrometric networks are extremely limited in spatial coverage and the quality of available records. Worldwide, there has also been a decline in both developed and developing countries in hydrometric data being collected (Rodda 1998; Sene & Farquharson 1998; Lanfear & Hirsch 1999; Hannah *et al.* 2011; Mlynowski *et al.* 2011). Even the UK, which is a well-monitored country with one of the longest and densest meteorological and hydrometric records worldwide (Dixon *et al.* 2013), has inevitable gaps in the hydrometric network and areas that are ungauged. Likewise, the uncertainty in flow estimation is seldom considered in practical applications prompting calls for improved engagement with the issue (McMillan *et al.* 2012; Coxon *et al.* 2015; Kiang *et al.* 2018). For example, the use of rating curves to stage measurements is by far the most common way to indirectly estimate the existing streamflow. Even in the best conditions, there will always be some uncertainty in the observed discharge. It derives from the accuracy of the instrumentation (especially for low flows), the changing nature of the flow regime (e.g. hysteresis, roughness variation and drowning conditions), the suitability of the mathematical expression being fitted, or the need to extrapolate the curve out of the range of measurements (McMillan *et al.* 2012). This implies that, on many occasions, available hydrometric records in a certain

doi: 10.2166/nh.2019.006

catchment may be inconsistent. The same is true when calculating the catchment average rainfall based on a certain interpolation technique and a number of rain gauges with potential measurement problems too (Neff 1977; Moulin *et al.* 2009), or when estimating the potential evapotranspiration (PET) based on sparse weather information and uncalibrated techniques (Andréassian *et al.* 2004). Notwithstanding the long-term need to improve meteorological and hydrometric network coverage and hydrological practices, hydrologists are faced with the challenge of estimating flows in ungauged or partially gauged basins or with data of unknown quality.

The challenge of estimating flows in ungauged basins has received significant attention in the last decade among hydrologists (see the Predictions in Ungauged Basins initiative in Sivapalan *et al.* (2003) and Hrachowitz *et al.* (2013)). Current methodologies can be classified into (1) regionalisation approaches (Wagener *et al.* 2004) designed to transfer the parameters of a hydrological model calibrated in gauged basins to ungauged basins by establishing a regression between them and catchment descriptors (He *et al.* 2011) and (2) response signature methods, based on a regression between hydrological signatures and catchment descriptors obtained from gauged basins, and used to condition the predictions generated by a hydrological model within a Bayesian framework (Holmes *et al.* 2002; Young *et al.* 2003; Bulygina *et al.* 2009, 2011; Wagener & Montanari 2011; Almeida *et al.* 2015). Fully distributed hydrological models, which explicitly consider the spatial distribution of physical properties across the catchment, when applied to estimate flows at ungauged locations, can be assigned to the first category, providing that realistic relationships between model parameters and spatial properties are adopted (Refsgaard 1997). This kind of approach has been criticised for its excessive determinism (Beven 2001), but when model parameters are obtained from physical properties through multiple linear regressions, the uncertainty can be significantly reduced (Blöschl 2005). While this technique has successfully been applied to many cases (Reed *et al.* 2004), the widespread use of distributed models in the industry is limited due to their relative complexity, something which this paper seeks to challenge.

However, distributed models can also offer an effective tool to identify potential problems in both input data and the observed flows used to validate a hydrological model. While the initial emphasis as regards hydrological uncertainty was placed on parameter equifinality (Beven & Binley 1992; Freer *et al.* 1996; Vrugt *et al.* 2003), the relevant role input and output uncertainty can play in the total runoff estimation error has recently received significant attention, usually involving a Bayesian approach (Vrugt *et al.* 2009; Renard *et al.* 2011; Le Coz *et al.* 2014; Coxon *et al.* 2015). Nevertheless, in applied hydrology, the uncertainty in input and output data is often not considered, being in best cases limited to a sensitivity test of assumptions. As flow series are often used in subsequent modelling exercises to optimise the operation of the current infrastructure or to select the most suitable new infrastructure, dealing with hydrological uncertainty can be cumbersome. Therefore, if practitioners assume that existing measurements are true representations of the variable of interest, there is an increased chance of introducing a bias in the analysis. In this context, distributed models can provide consistency in evaluating the goodness of observations. Assuming that the model is capable of sufficiently representing the physical processes involved in the flow generation when model parameters are derived across the basin based on similar spatial information, and with the same degree of accuracy and quality, deviations between modelled and observed flows can be attributed to deficiencies in data. Likewise, when data have been verified, deviations can be due to the existence of a particular mechanism not adequately captured by the model. This use of models as scientific tools to test data or hypotheses is far less frequent than the most widely used application for scenarios simulation (Silberstein 2006).

Conversely, lumped models, with a poor connection to physical properties, are likely to ignore these deficiencies and therefore risk being able to reasonably simulate observed flows despite problems in the source information. Parameters may then be biased and their application to other periods, historical or future, can be imperfect (Wilby 2005; Oudin *et al.* 2006; McMillan *et al.* 2010). In heavily modified basins, where the extent of artificial influences is large, the existence of uncontrolled or poorly monitored abstractions/returns in certain parts of the basin will go unnoticed in lumped models but can be inferred from a distributed analysis (Nalbantis *et al.* 2011). Notwithstanding this, lumped models are still predominant in hydrological practice given the quickness and easiness of its application and, therefore, their related lower costs.

In addition to the primary purpose of the work to assess water resource yields, this paper aims to demonstrate the use of fully distributed modelling as a catchment-wide data validation tool in three catchments in the East of England with data coverage and quality issues. Specific objectives were to overcome the limitations in the hydrometric station coverage and critically review the quality and assumptions behind the input data. In particular, artificial influence information is known to be incomplete across the three catchments, as in addition to significant licenced abstractions there are unmonitored transfers, not requiring a licence, to lowland internal drainage board (IDB) areas (National Rivers Authority 1993). A previous hydrological modelling study within the part of the basin (Mott MacDonald 2007) adopted a semi-distributed model. Although consideration was given to adopting a semi-distributed model in this study, this was not considered a viable option because of the gaps in gauging station coverage and quality and the resulting significant scaling for ungauged areas. As such, it was considered that a distributed modelling approach would provide improved basin representation.

In the case of the Ely Ouse, since groundwater interactions are important, consideration was given to adopting the Environment Agency's North East Anglia Chalk (NEAC) model (Black *et al.* 2012). However, the purpose of the model is primarily to understand the impacts of groundwater abstraction, and as a tool to regulate groundwater abstraction licenses, and as such, this model was not readily applicable to surface water applications.

In the specific context of these catchments, accuracy was required for the derivation of flow estimates, as the results were used directly to facilitate a water company water resources assessment of the East of England. The wider importance of this cannot be underestimated, as all water companies are responsible for maintaining public water supply and analysing options to ensure this supply remains secure in the face of growing demand and increasing environmental pressures. This implies a need to establish reliable baseline flow series across catchments to allow the understanding of current and future water availability and to inform the development and appraisal of possible future water supply options. A small deviation in simulated flows might modify the feasibility of certain proposed infrastructure to a point where another alternative would be preferred, with potential monetary consequences for the final consumer. This issue is not only relevant in the UK but in any part of the world regardless of the temporal and spatial coverage of meteorological and hydrometric information. Having a clear understanding of the reliability of input data and deriving a good basin representation are essential when attempting to quantify available flow for potential abstraction for any water supply project (such as domestic, agricultural, industrial or hydropower).

Study area

With approximately 600 mm of annual rainfall and less than 200 mm of effective annual rainfall, the East of England is the driest region of the UK, which, with the population and agricultural demand, results in the existence of serious water stress (Environment Agency and National Resources Wales 2013). This situation is expected to worsen given the ongoing demographic expansion and the impact of climate change (Christierson *et al.* 2012; von Lany *et al.* 2013). In addition, new sources of cheap and abundant water are scarce, and securing the future supply requires a difficult exercise of finding an optimal solution. An accurate assessment of water availability is therefore critical. The three catchments under specific consideration as part of this study are the Ely Ouse, Witham and Black Sluice (see Figure 1).

The Ely Ouse catchment drains an area of 3,581 km^2 upstream of the Denver Complex, including the major tributaries of the Rivers Cam, Lark, Little Ouse and Wissey. The tributaries drain the uplands of the Great Ouse Chalk outcrop and have a heavily modified hydrological regime. There are no major reservoirs in the studied area. However, there are numerous abstractions (including the diversion of water to South Essex through the Ely Ouse to Essex Transfer Scheme) and effluent discharges that, due to their magnitude, substantially alter the natural flow regime. The River Witham catchment drains an area of 2,066 km^2, including tributaries of the Rivers Brant, Bain, Slea and Barlings Eau which drain the Lincolnshire Limestone and Spilsby Sandstone outcrops. Similarly, this catchment also has numerous abstractions and also the

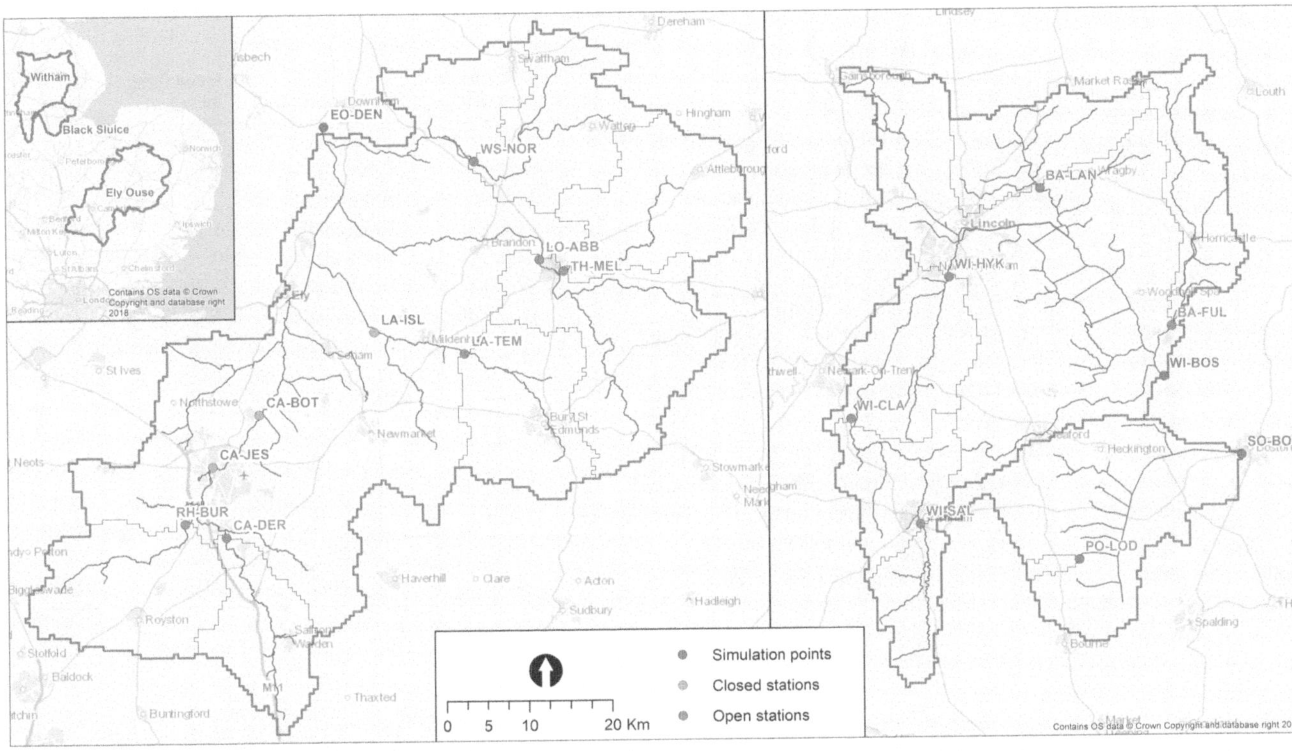

Figure 1 | Location of gauging stations used in the study.

Trent-Witham-Ancholme Transfer Scheme, which serves to supplement flows in the River Witham (Environment Agency 2013b). The Black Sluice catchment, which is drained by the South Forty Foot Drain, has an area of 520 km^2 covering high-grade agricultural parts of the low-lying Lincolnshire Fens.

All three catchments under consideration are predominantly rural and drain towards the Wash, and have lower sections that have been extensively drained and managed by a system of drains and dykes, allowing the high-grade low-lying agricultural peatland to be farmed (Environment Agency 2013a). The Denver Complex, comprising several sluices and gates, helps to control and manage water in the Ely Ouse catchment, allowing the protection of low-lying fenland from tidal flooding and the drawing of river flows into low-lying drains via slackers for crop irrigation (Environment Agency 2013a). The Grand Sluice and Boston Lock serve similar functions for the River Witham and South Forty Foot Drain, respectively.

The Ely Ouse and Witham catchments are seemingly well instrumented with numerous gauging stations as summarised in Table 1 and shown in Figure 1 (considering drainage areas greater than 200 km^2 for the Ely Ouse and greater than 100 km^2 for the Witham and any gauging station for the Black Sluice catchment). In the case of the Ely Ouse, there has been a decline in the station coverage in downstream areas in recent years.

However, the River Witham lacks a main gauging station on its lower section, and the one covering the largest drainage area (North Hykeham, which was put in 1998) was thought by the operator (Environment Agency) to underestimate flows, and there is not yet consensus about its validity (Dan Burbidge, personal communication, 20 July 2017). This is due to an apparent much lower runoff than at the upstream gauging station at Claypole Mill and the lack of a proper control structure to measure low flows. Likewise, the most downstream station for the Ely Ouse, at Denver Complex, contains major gaps and inconsistencies (National River Flow Archive 2017) as it is a composition of the flows through five sets of sluices whose value is obtained from hydraulic calculations. A review and improvement of flow derivation methods at the Denver Complex (Environment Agency 2012) increased the accuracy of flow records from 2009 onwards, although there are still some concerns with

Table 1 | Gauging stations used in the study

Basin	Official code	Model code	Name	Catchment area (km²)	Period of record	Mean flow (m³/s)	Q95 flow (m³/s)
Ely Ouse	33035	EO-DEN	Ely Ouse at Denver Complex	3580.7	1958–2015	15.79	1.80
	33003	CA-BOT	Cam at Bottisham	807.3	1936–1987	3.66	0.91
	33016	CA-JES	Cam at Jesus Lock	769.1	1959–1983	2.81	0.71
	33021	RH-BUR	Rhee at Burnt Mill	306.1	1962–2015	1.18	0.26
	33024	CA-DER	Cam at Dernford	199.6	1949–2015	0.92	0.27
	33004	LA-ISL	Lark at Isleham	464.4	1936–1986	1.80	0.44
	33014	LA-TEM	Lark at Temple	274.1	1960–2015	1.29	0.46
	33034	LO-ABB	Little Ouse at Abbey Heath	707.7	1968–2015	3.83	1.16
	33019	TH-MEL	Thet at Melford Bridge	311.4	1960–2015	1.89	0.46
	33006	WS-NOR	Wissey at Northwold Total	259.4	1956–2015	1.88	0.47
Witham	30001	WI-CLA	Witham at Claypole Mill	297.9	1959–2015	1.88	1.26
	30005	WI-SAL	Witham at Saltersford Total	126.1	1988–2015	0.82	1.17
	N/A	WI-HYK	Witham at North Hykeham	550.2	1998–2015	2.89	1.02
	30002	BA-LAN	Barlings Eau at Langworth Bridge	210.1	1960–2015	1.45	0.22
	30003	BA-FUL	Bain at Fulsby Lock	197.1	1962–2015	1.27	0.59
Black Sluice	30014	PO-LOD	Pointon Lode at Pointon	11.9	1972–2015	0.07	0.00

Source: National River Flow Archive (2017) and Environment Agency (2017).

the data quality. For these reasons, direct calibration against this data was not undertaken, and it was only used for model comparison purposes. The upstream parts of both catchments have gauged catchments with known differences in terms of meteorology, land use, soils and geology, which would prevent a reliable transposition downstream using traditional lumped models. For the Black Sluice catchment, the only available gauging station is located on the small Pointon Lode stream, covering only 12 km^2. Overall, the Ely Ouse, River Witham and Black Sluice catchments have, respectively, 49%, 46% and 2% of their areas effectively monitored by currently operational long-term gauging stations.

METHODOLOGY

Selected model and input data

The TETIS model, developed by the Research Group of Hydrological and Environmental Modelling (GIMHA) of the Technical University of Valencia in Spain, has been adopted to simulate the hydrological response of the catchments, making it the first application in the UK. For a detailed description of the TETIS model, including a conceptual schematic, see Francés *et al.* (2007) and UPV-GIMHA (2014). The TETIS model is a fully distributed rainfall-runoff algorithm, in which the catchment is divided into a number of cells whose physical properties are assumed homogeneous and characterised by a series of tanks linked vertically and representing different hydrological processes (discussed further below). In each cell, these hydrological processes are represented by linearised approximations of the non-linear differential equations that govern the movement of water. In this sense, TETIS can be classified as a conceptual model, but with parameters that are assigned to physical properties. All cells drain downstream according to their elevation until the flow reaches the river channel, where the flow is routed to the model outfall. Two model domains were set up: one for the Ely Ouse catchment and one covering both the Witham and Black Sluice catchments. A 500 m cell size grid was adopted as a compromise between computation requirements and the ability to sufficiently capture the spatial variability.

National 5 km by 5 km gridded rainfall (Met Office) and PET (MORECS v2.0) datasets (Hough & Jones 1997) were adopted as input data for the two models. Major public water supply abstractions and effluent discharges were introduced at individual locations where possible. Other smaller abstractions (mainly for agricultural purposes) were included in an aggregated way based on data provided by the Environment Agency and assigned to the cell upstream of the respective gauging station or the point of interest. For the Ely Ouse catchment, groundwater augmentation schemes were also modelled; based on advice from the Environment Agency, a proportion of the augmentation was assumed to be lost due to infiltration. For the Witham, the Trent-Witham-Ancholme Scheme was included introducing an additional inflow at Lincoln. In order to avoid overcounting the stormwater captured by the sewerage network, the monthly Q80 value (i.e. the flow rate exceeded with a probability value of 80% during each month) was estimated as representative of the Dry Weather Flow for effluent discharges.

Model parameters

Water balance and fluxes at each cell are controlled by 11 parameters (numbered below and taking the form of GIS raster layers within the model), which reflect spatial variability across the catchment. These GIS raster layers are derived from physical properties in a uniform way across the catchment to enable the transfer of information attained at gauged stations to ungauged locations (Götzinger & Bárdossy 2007). The parameters are modified by nine global correction factors (two parameters, the interception capacity and the soil optimum water content, are fixed) that are adjusted during calibration and scale the GIS raster layers consistently.

An Ordnance Survey 50 m Digital Elevation Model, resampled to 500×500 m cells, and conditioned to reflect the river network, was used to derive flow direction, accumulated drainage area and slope. The latter was used to characterise the overland flow velocity v (1) at each cell (see Figure 2(a)) and for in-channel flow routing. Maximum interception capacity $Imax$ (2) was related to vegetation cover as well as the crop coefficient λ (3) that modifies the PET through the year. Vegetation cover throughout the studied region was defined based on the Corine Land Cover 2006 dataset (European Environment Agency 2007).

Overland flow can occur due to both soil saturation (the Dunnean mechanism) and the exceedance of the infiltration capacity (the Hortonian mechanism). Evapotranspiration depends on the type of vegetation cover and varies seasonally, being also reduced as the soil approaches the wilting point. Overland flow, interflow and baseflow are modelled with a linear reservoir approach, with constants derived from the expected velocity over the surface or through the soil. Flow is routed through the river channel by a kinematic wave, in which the size and roughness of the channel are a function of the contributing drainage area (Bérod et al. 1999).

The static storage capacity H_u (4), which represents the maximum water that can be held in each cell without generating flow, comprises two components: surface detention capacity, which is a function of the vegetation cover and the surface slope, and soil tension capacity, which is the water stored in the soil matrix by capillary forces and only used by plants for their transpiration. The soil tension capacity was estimated as the product of the soil thickness and the difference between the field capacity and the wilting point. The HORIZON Hydraulics version 2 dataset (Hollis et al. 2015), developed by the Cranfield University Soil and AgriFood Institute (CSAI), provides an estimation of this information for each soil horizon. In the case of urban areas, it was limited to the surface storage capacity (see Figure 2(b)). The soil optimum water content θ (5), below which the PET is reduced due to water stress, was established as 60% of the tension storage. Likewise, the maximum capacity of the gravitational tank H_g (6) can be defined as a certain percentage of the static storage capacity. When exceeded, water can move to the surface and produce overland flow. A similar process was conducted to characterise both the infiltration capacity k_s (7) and interflow velocity k_{ss} (8), which were associated with the sub-vertical hydraulic conductivity of the topsoil and the weighted averaged horizontal hydraulic conductivity of the whole soil package, respectively (see Figure 2(c) and 2(d)). Further spatial variability was introduced by considering the soil texture. Several studies conducted by the European Soil Data Centre were used to characterise the hydraulic properties of the soil based on the abundance of sand, clay and organic matter (de Brogniez et al. 2015; Ballabio et al. 2016) and by applying pedotransfer functions fitted to a wide range of soils around the world (Saxton & Rawls 2006).

The percolation capacity k_p (9) into the aquifer and the connected aquifer flow velocity k_{sa} (10) are directly related

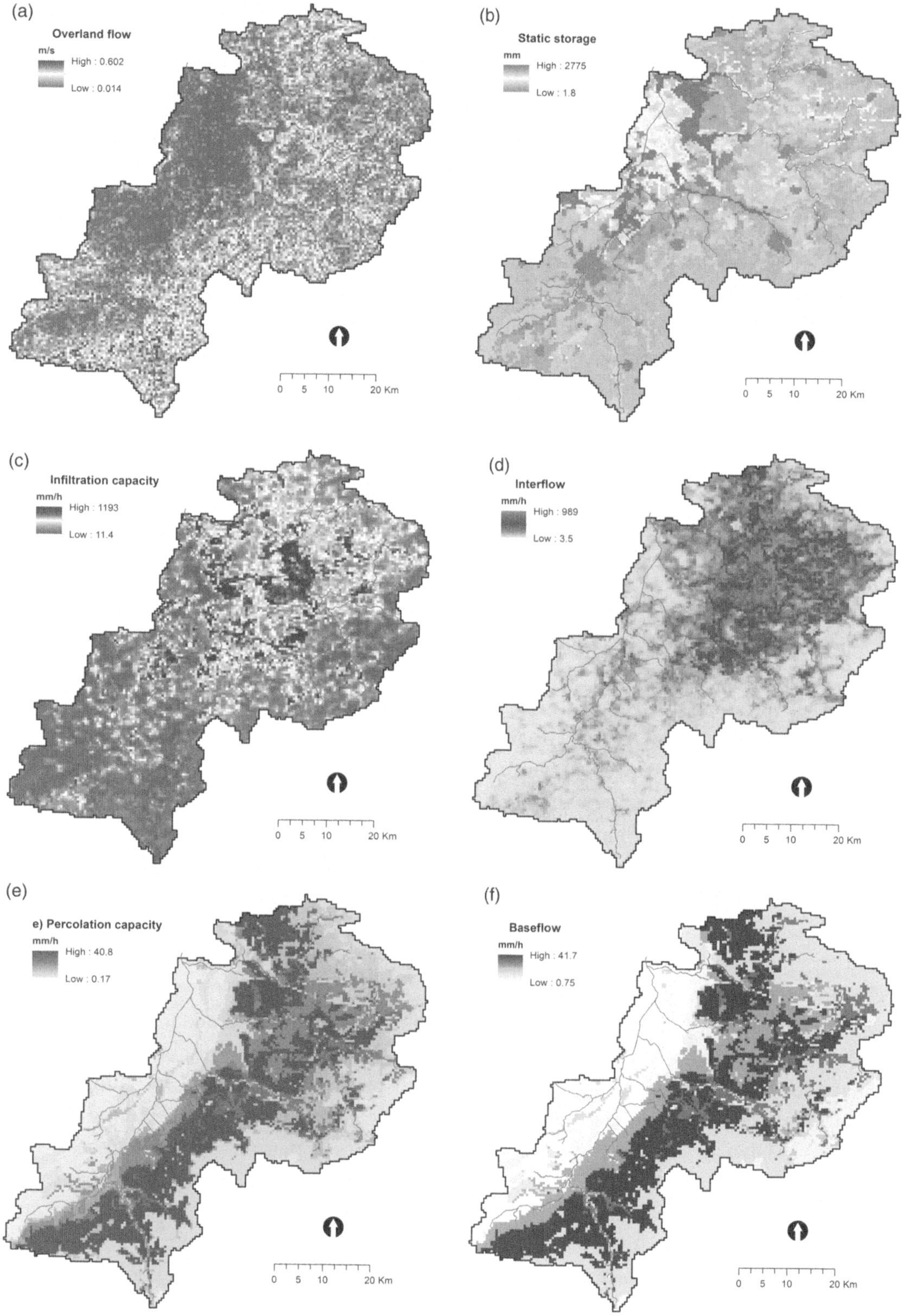

Figure 2 | Spatial distribution of main model parameters for the Ely Ouse: (a) overland flow velocity, (b) static storage capacity, (c) infiltration capacity, (d) interflow velocity, (e) percolation capacity and (f) baseflow velocity.

to the permeability of the bedrock and superficial deposits. Maximum and minimum bedrock and superficial deposit permeability classes obtained from the British Geological Survey 1:50,000 Digital Geological Map of Great Britain (Lewis *et al.* 2006) were used to establish a relative spatial distribution of aquifer properties by assigning a notional hydraulic conductivity to each class using a logarithmic scale (see Figure 2(e)). To account for the additional influence of soil thickness and texture on percolation, the initial aquifer flow velocity estimation was further modified with the Base Flow Index (the ratio of base flow to total flow) estimated for each soil association by the CSAI (see Figure 2(f)).

Finally, the deep aquifer percolation rate k_{ps} (11) refers to losses due to the lack of connection between the deep aquifer and the river network within the studied catchment. It has only been applied in the special case of the Ely Ouse catchment which contains a major aquifer in the Chalk. The Chalk is a thick cretaceous white limestone aquifer with low matrix permeability but a well-developed and interconnected network of fractures and is a major source of water supply in SE England (Allen *et al.* 1997). The behaviour of the Chalk is quite complex and highly variable. This complexity is also reflected in the unsaturated zone, which controls the recharge of the aquifer. Ireson *et al.* (2009) demonstrated that percolation from the surface could last several years during which water would not be available for generating baseflow. Isotopic analyses confirmed this fact, extending the recharge duration to several decades in some areas (Darling & Bath 1988). Given this, and in order to close the water balance, the activation of the deep aquifer percolation parameter is required, assuming a value of 1 for areas where the Chalk outcrop is present.

Calibration and validation

Even within the field of distributed hydrological modelling, parsimony is a desirable principle. Reducing the number of parameters increases the chances of their identifiability, allowing a more reliable temporal and spatial extrapolation (Wagener *et al.* 2002), and reducing parameter equifinality (Beven & Binley 1992). Therefore, model parameters should be calibrated regionally for the whole catchment so that the spatial relationships derived from physical properties are retained. This requires the use of correction factors that modify model parameters globally and that are calculated by comparing simulated and observed flows at one key gauging station in the catchment (Francés *et al.* 2007). As part of the calibration process, all individual cells for each parameter are scaled by the same equivalent global correction factor. In TETIS, there are nine correction factors that modify the storage capacities and the velocities of the different flow components (correction factors and other model start conditions are summarised in the Supplementary Material).

For the Ely Ouse model, the Little Ouse at Abbey Heath gauging station was selected for calibration for the period 1995–2014 due to it having relatively minor artificial influences, a reasonable gauged record and the most significant drainage area of the available stations. For the Witham and Black Sluice model, the Claypole Mill gauging station was selected for calibration for the period 1995–2014.

A four-step calibration process was adopted which involved an initial manual calibration of the water balance, followed by the application of the SCE-UA optimisation algorithm (Duan *et al.* 1994), after which two sets of further manual calibrations were made to firstly match the hydrograph shapes, followed by a verification of the water balance. The Nash–Sutcliffe coefficient (NSCC) (Nash & Sutcliffe 1970) and the volume error were both adopted as objective functions with the target of achieving a mean flow percentage error of less than 5% and an NSCC greater than 0.7 for calibration. Further details on the calibration strategy are contained in the Supplementary Material. A subsequent validation process was undertaken for other gauging stations in the basin with the aim of verifying the transferability of the calibration (Vélez *et al.* 2009) or to explore a potential inconsistency.

RESULTS

Calibration performance

Figure 3 shows the comparison of calibration flow duration curves (FDCs) achieved for the Little Ouse at Abbey Heath and River Witham at Claypole Mill which shows a good resemblance. Calibration statistics (see Table 2) were

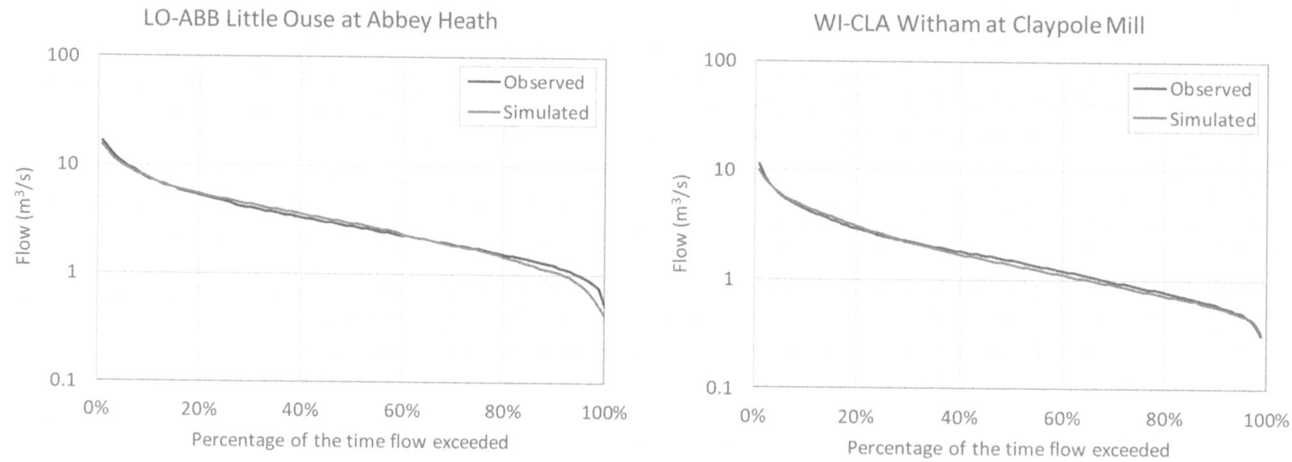

Figure 3 | Little Ouse at Abbey Heath and River Witham at Claypole Mill observed and simulated calibration FDCs (1995–2014).

Table 2 | Statistics of the calibration (bold) and the validation process

Model	Gauging station	Period for statistics	Mean flow percentage error	NSCC	*P* (mm)	PET (mm)	Runoff observed (mm)	Runoff simulated (mm)
Ely Ouse	**Little Ouse at Abbey Heath**	**1995–2014**	**− 0.4%**	**0.85**	**637**	**648**	**167**	**168**
	Rhee at Burnt Mill	1995–2014	0.4%	0.81	587	651	115	116
	Cam at Dernford	1995–2014	11.2%	0.70	619	650	135	150
	Lark at Temple	1995–2014	5.0%	0.73	631	644	149	157
	Thet at Melford Bridge	1995–2014	−14.0%	0.82	666	670	204	175
	Wissey at Northwold Total	1995–2014	−22.1%	0.58	700	670	237	184
	Ely Ouse at Denver	1995–2014	–	–	622	656	–	120
		2009–2014	0.4%	0.80	602	728	136	137
Witham and Black Sluice	**Witham at Claypole Mill**	**1995–2014**	**− 1.9%**	**0.82**	**637**	**662**	**226**	**223**
	Witham at Saltersford Total	1996–2014	5.6%	0.79	686	658	215	228
	Witham at North Hykeham	1999–2014	11.5%	0.71	645	664	180	200
	Barlings Eau at Langworth Bridge	1995–2014	−12.5%	0.65	664	667	248	217
	Bain at Fulsby Lock	1995–2014	−1.4%	0.80	721	664	258	256
	Pointon Lode at Pointon	1995–2014	−9.8%	0.58	648	665	201	183

considered good with both models achieving an NSCC greater than 0.8 and a match to the mean flow within 2%. Comparison of the hydrographs is included in the Supplementary Material.

Validation performance

The correction factors established in the calibration process have been applied to the whole catchment, with flows simulated at the locations of the remaining gauging stations and compared to the observed record as a spatial validation of the model performance. If the calibrated model predicts the observed flows at these gauging stations well, the reliability of simulated flows at ungauged locations is likely to be high as long as the records for the validation stations are considered accurate. Deviations between modelled and observed flows at validation stations can also be indicative of measurement problems. In this regard, Table 2 also includes the results of the comparison between simulated and observed flows for validation stations. For some stations, the reference period had to be restricted due to the availability of data, particularly for the Ely Ouse at Denver which only had suitable reference data for the period 2009–2014.

Likewise, a comparison of FDCs for the corresponding validation periods listed in Table 2 has been included in the Supplementary Material. As can be seen from Table 2, the distributed modelling approach is, in general, able to

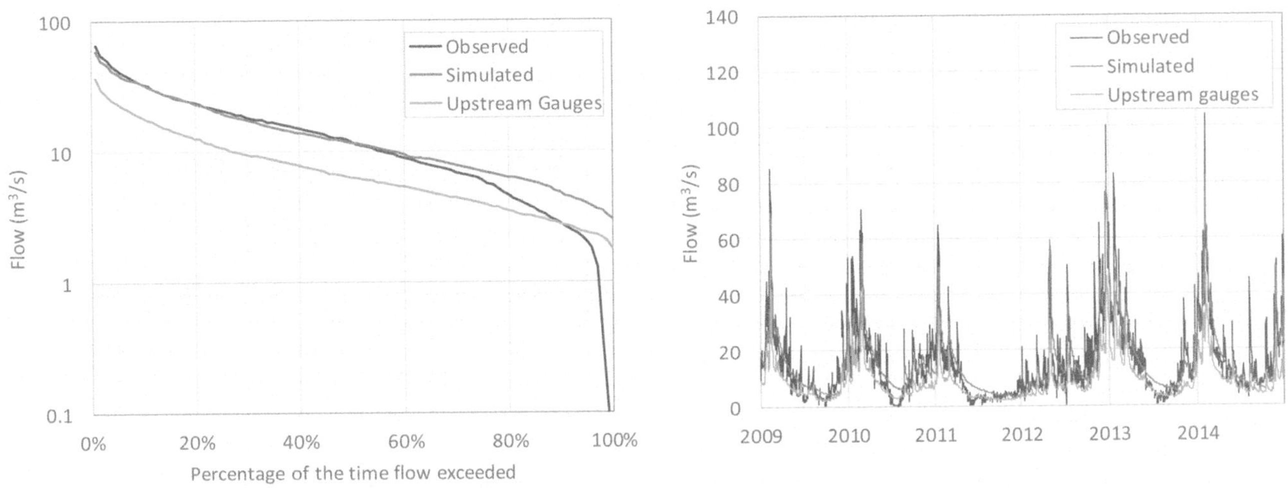

Figure 4 | Ely Ouse at Denver FDC (left) and daily flow series (right) (2009–2014).

provide a good representation of flows at the validation stations, with some achieving comparable statistical performance to the calibration stations.

Verification of meteorological and hydrometric data

Various stations used in the model validation were subject to concerns regarding the quality of the flow record available. Given the availability of a limited validation flow record for the Ely Ouse at Denver for the period 2009–2014, specific mention of this location is warranted.

A comparison of the observed and simulated FDC (Figure 4) shows a good match across the upper half of the flow range, with an increasing deterioration towards the lower end. The daily flow series shows that there are significant fluctuations and sudden drops in flows that are likely to be associated with water management activities in the low-lying parts of the catchment, a feature which is not present when summing the flows from the upstream gauged stations (see Figure 1). The fact that at times, there is less water available at Denver than at the sum of the upstream gauged stations is an indication of unrecorded transfers from the Ely Ouse into low-lying IDB areas via slackers for water-level management purposes. This is a long-standing arrangement that pre-dates licencing, which in the past has been estimated to account for 50% of river flow from the upland area (National Rivers Authority 1993). In drier periods, IDBs actively retain water to maintain higher levels in drains and channels so that water conveyance is ensured and ecosystems preserved (Hodge & McNally 2000). Part of this water infiltrates back into the soil enhancing evapotranspiration and effectively acting as irrigation. River bed and bank leakage may also be a factor in contributing to channel flow loss. Despite these issues at lower flows, the broad visual fit between the simulated and observed series and a reasonable NSCC gives confidence that the modelling approach delivers improved representation of these ungauged areas.

The model has also highlighted a high volume error and a low NSCC for the Wissey at Northwold. The examination of the daily flow series indicates that while there is a reasonable match against the observed flows during the first half of the period, the model increasingly fails to replicate them in the latter half (see Figure 8 in the Supplementary Material). While uncertainties in input datasets cannot be ruled out, since the station record (National River Flow Archive 2017) cites regular drowning of the weir as an issue and a lack of adjustment in periods of anomalously high flows, an inaccurate record is likely to be contributing to the high volume error. Similarly, the Thet at Melford Bridge weir is subject to drowning due to downstream weed growth (National River Flow Archive 2017), which is likely to explain the relatively high volume error, particularly given that at the calibration station the volume error is minimal.

The reliability of low flow measurements for Pointon Lode in the Black Sluice catchment is also considered uncertain as it is known to drown at high flows and the record displays some anomalous recessions, for example in

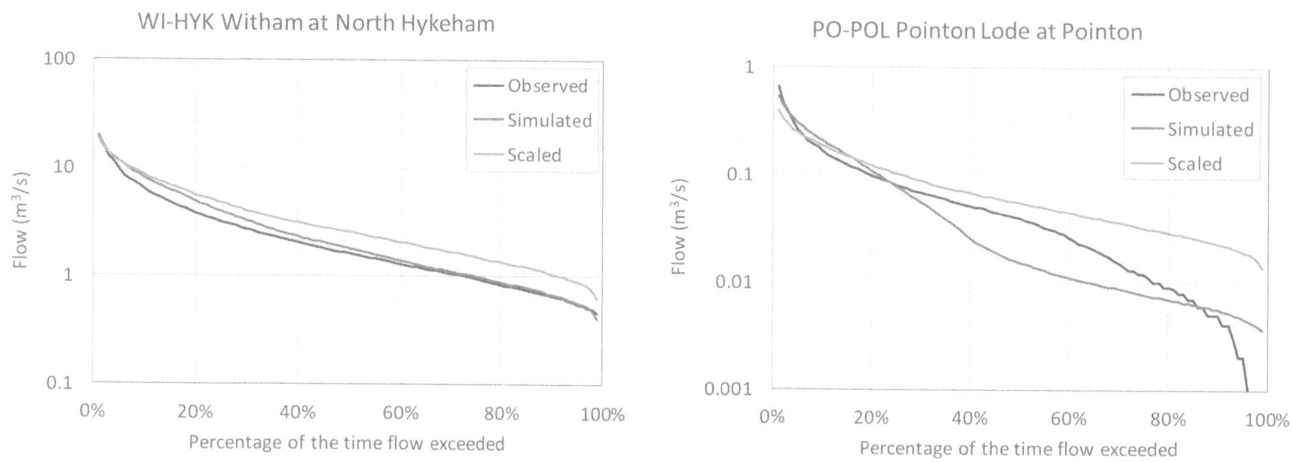

Figure 5 | Witham at North Hykeham FDC (left) and Pointon Lode at Pointon FDC (right) (1995–2014). Green line represents the Claypole Mill FDC scaled as a function of drainage area.

1994, 1998 and 1999, the reason for which is unknown (National River Flow Archive 2017). Under these circumstances, the model would be expected to undersimulate against the record, as can be seen in Figure 5. As an additional verification of the simulation in the Black Sluice catchment, results from the water balance assessment undertaken within the Water Resources East project (Mott MacDonald 2017) to determine the contribution of the Lincolnshire Limestones to the South Forty Foot Drain were compared with the outputs of the TETIS model at 10 simulation points along the springline villages, yielding only a 3% difference. In addition, the estimated average flow passing the downstream pumping station, just west of Boston, was only 1% lower than the value simulated with the calibrated model.

Likewise, flows at North Hykeham in the Witham have historically been thought to be underestimated (Dan Burbidge, personal communication, 20 July 2017). In fact, the available spot flow measurements point towards an underestimation of 5% in low flows (below 1 m^3/s) and of 10% in the rest of the flow range. After correcting the observed flow series due to this bias, the model was able to provide a very good match to the medium and low flows (see Figure 5), much better than if the FDC at the upstream Claypole Mill gauging station was scaled only as a function of drainage areas to each station, something that would have been done if a lumped model would have been adopted and calibrated at Claypole Mill. This shows that changes in rainfall and physical properties in the intermediate catchment are properly captured by the model, and reinforces the need for updating the North Hykeham rating curve based on the spot flow measurements.

DISCUSSION

The modelling results have clearly shown an improved representation of ungauged areas when compared to scaling as a function of the drainage area, which would be similar to the procedure if a lumped model would have been adopted. In many applications, this reason alone would be enough to justify the adoption of a distributed modelling approach. However, this application of distributed modelling has highlighted wider benefits associated with the verification of hydrological data and improved catchment understanding.

The flagging of potential bias in meteorological datasets commonly used by the hydrological community, either highlighting a problem in the methodology or in the raw data coming from certain stations, is a key benefit. Without this approach highlighting water balance discrepancies across the area, it may not have been identified that the MORECS v2.0 PET dataset (Hough & Jones 1997) was derived without using any climatological stations within the Ely Ouse basin. The resultant interpolation of climatological variables may therefore not reliably reflect local conditions. MORECS v1.0 (pre-1996) has much lower variation in PET across the catchment, which in addition to the above factor casts some doubt on the validity of the MORECS v2.0 PET dataset for the basin. It should be noted that this issue does not appear to affect the Witham-Black Sluice model as there is lower variability in PET (Table 2). Highlighting these issues can help in the update and improvement of these meteorological datasets.

Other spatial differences in calibration performance highlighted by using distributed modelling can help point to potential model improvements in the representation of hydrological processes in different areas. A particular issue for the Ely Ouse catchment is groundwater influencing baseflows. Water percolating to the Chalk aquifer, which stretches across the central part of the catchment, can return to the river network either at springs or gradually along lengths of the river (Environment Agency 2013a). The validation gauging stations are all located within the Chalk and will, to varying degrees, be affected by losing to, or gaining from, the aquifer. While deep aquifer losses have been simulated in the model for the Chalk outcrop, the current configuration of TETIS has not allowed simulation of groundwater returns to the channel. This may explain the inability to obtain a precise water balance for the smaller upstream catchments (e.g. the Cam at Dernford).

The issue of unrecorded transfers has been readily identified using distributed modelling. In the case of the Ely Ouse, this is potentially 2 m^3/s at Q95. In the Witham catchment, there is also a concern that this may be a contributory factor to the weaker model performance for the Barlings Eau at Langworth, a catchment which is known to be affected by irrigation abstractions. The verification of these issues could allow the water regulator to better understand the scale of these transfers and the overall impact from both licenced and non-quantified permitted abstractions.

As outlined in the results, there are multiple examples where the model has highlighted issues with hydrometric station performance. This can help focus attention on the gauging stations that require improvement. This could lead to the development of an enhanced rating curve based on new spot flow gaugings and a detailed hydraulic model, or alternative methods for the measurement of non-modular flow (if backwater effects or seasonal changes cannot be avoided), or relocation of the station.

A better understanding of all these issues, thanks to the use of a distributed approach, has several benefits. In summary, distributed modelling could help improve the quality of the hydrological information, which is the basis of any water resources management or flood-risk assessment, leading to the best informed planning processes.

The progressive increase in publicly available spatial datasets characterising the physical properties of the region obtained through remote sensing is now enabling the application of this kind of approach worldwide, even in remote areas. With reducing related costs, GIS efficiencies and model run automation, it is increasingly possible to deliver distributed modelling as a commercially and practically viable option. Distributed modelling can open opportunities to enhance hydrological assessments in the industry, particularly in data scarce and uncertain environments.

CONCLUSIONS

Fully distributed modelling has been applied as part of the overarching purpose of assessing water resource yields but has also served as a valuable catchment-wide data validation tool in three catchments in the East of England with data coverage and quality issues. The approach has successfully shown its capability of considering differences in climate and physical properties when transposing the calibration made at a key gauging station to other points, maximising the potential of limited available flow data within the area and providing improved basin representation compared to traditional lumped modelling approaches.

Moreover, it has been demonstrated how the models can be used to identify potential deficiencies in the reference flow data across all considered catchments. This is a valuable tool to supplement gauging station reviews and has significant potential to help deliver improved modelling results in situations where gauged flow data are of variable quality. This can be particularly relevant when investment decisions (e.g. the size of a reservoir or the location of a river abstraction) are to be made based on the availability of water resources at either an ungauged location or a location where the existing hydrometric data are dubious. It is acknowledged that distributed modelling is more costly than lumped modelling due to its increased complexity. However, if simpler lumped modelling approaches are not adequate to capture the catchment characteristics or resolve issues with data coverage and quality, the additional costs of the distributed approach can be small in comparison with the investment decisions being made.

The models have also identified potential issues with other input datasets, in this case the MORECS v2.0 PET dataset and transfers for irrigation abstractions. The wider use of distributed modelling within the water resources industry will only help with the identification and feedback of these issues to data custodians and water regulators and help to improve the accuracy and reliability of water resource assessment in the long term. The ease of identification of these issues presents an advantage over lumped or semi-distributed modelling approaches, where potential errors may be masked by the calibration of the input parameters and be more difficult to identify. While input uncertainty and output uncertainty are not explicitly defined, the approach reduces uncertainty as it takes into account spatial consistency, leading to more robust predictions.

The study has also highlighted areas for further work and investigation. Unrecorded transfers (pre-dating licencing requirements) are a specific issue in this area that despite being a known long-term issue still requires further consideration and quantification to improve the model reliability for low flows. The representation of groundwater processes within the TETIS model in complex chalk streams is also an aspect that needs further development. While there are recommendations for further model refinement and improvement, the work has demonstrated the benefits of the application of distributed modelling. In the international context, where often even less reliable hydrological data are available, the potential benefits are even greater. The paper has also demonstrated that, with the progressive increase in publicly available spatial datasets, fully distributed hydrological modelling can be effectively and beneficially implemented in commercial water industry applications, similar to groundwater modelling. As such, a wider consideration of these hydrological approaches within the industry is strongly recommended.

ACKNOWLEDGEMENTS

The authors acknowledge the support of Anglian Water Services and Mott MacDonald in publishing this paper. Thanks are also due to Dan Burbidge (Environment Agency) for his useful advice regarding the gauging stations and artificial influences present in the catchments. We are also grateful to Peter Ede for his helpful comments on the manuscript.

SUPPLEMENTARY MATERIAL

The Supplementary Material for this paper is available online at https://dx.doi.org/10.2166/nh.2019.006.

REFERENCES

Allen, D. J., Brewerton, L. J., Coleby, L. M., Gibbs, B. R., Lewis, M. A., MacDonald, A. M., Wagstaff, S. J. & Williams, A. T. 1997 The physical properties of major aquifers in England and Wales. *British Geological Survey Technical Report, WD/97/34.* Environment Agency Research and Development Publication 8.

Almeida, S., Le Vine, N., McIntyre, N., Wagener, T. & Buytaert, W. 2015 Accounting for dependencies in regionalized signatures for predictions in ungauged catchments. *Hydrology and Earth System Sciences Discussions* **12** (6), 5389–5426. doi:10.5194/hessd-12-5389-2015.

Andréassian, V., Perrin, C. & Michel, C. 2004 Impact of imperfect potential evapotranspiration knowledge on the efficiency and parameters of watershed models. *Journal of Hydrology.* doi:10.1016/j.jhydrol.2003.09.030.

Ballabio, C., Panagos, P. & Monatanarella, L. 2016 Mapping topsoil physical properties at European scale using the LUCAS database. *Geoderma.* doi:10.1016/j.geoderma.2015.07.006.

Bérod, D. D., Singh, V. P. & Musy, A. 1999 A geomorphologic kinematic-wave (GKW) model for estimation of floods from small alpine watersheds. *Hydrological Processes.* doi:10.1002/(SICI)1099-1085(19990630)13:9<1391::AID-HYP809>3.0.CO;2-B.

Beven, K. 2001 How far can we go in distributed hydrological modelling? *Hydrology and Earth System Sciences* **5** (1), 1–12. doi:10.5194/hess-5-1-2001.

Beven, K. & Binley, A. 1992 The future of distributed models: model calibration and uncertainty prediction. *Hydrological Processes.* doi:10.1002/hyp.3360060305.

Black, A. D., Lewis, R. T., Grout, M. W. & Witterick, W. R. 2012 Crossing boundaries, the influence of groundwater model boundaries and a method to join and split MODFLOW models. *Geological Society, London, Special Publications* **364** (1), 155–172. doi:10.1144/SP364.11.

Blöschl, G. 2005 Rainfall-runoff modeling of ungauged catchments. In: *Encyclopedia of Hydrological Sciences* (Malcolm G. Anderson & Jeffrey J. McDonnell, eds). John Wiley & Sons, Ltd., Chichester, UK. doi:10.1002/0470848944.hsa140.

Bulygina, N., Mcintyre, N. & Wheater, H. 2009 Conditioning rainfall-runoff model parameters for ungauged catchments and land management impacts analysis. *Hydrology and Earth System Sciences.* doi:10.5194/hessd-6-1907-2009.

Bulygina, N., McIntyre, N. & Wheater, H. 2011 Bayesian conditioning of a rainfall-runoff model for predicting flows in ungauged catchments and under land use changes. *Water Resources Research*. doi:10.1029/2010WR009240.

Christierson, B. v., Vidal, J. P. & Wade, S. D. 2012 Using UKCP09 probabilistic climate information for UK water resource planning. *Journal of Hydrology*. doi:10.1016/j.jhydrol.2011.12.020.

Coxon, G., Freer, J., Westerberg, I. K., Wagener, T., Woods, R. & Smith, P. J. 2015 A novel framework for discharge uncertainty quantification applied to 500 UK gauging stations. *Water Resources Research*. doi:10.1002/2014WR016532.

Darling, W. G. & Bath, A. H. 1988 A stable isotope study of recharge processes in the English Chalk. *Journal of Hydrology*. doi:10.1016/0022-1694(88)90026-1.

de Brogniez, D., Ballabio, C., Stevens, A., Jones, R. J. A., Montanarella, L. & van Wesemael, B. 2015 A map of the topsoil organic carbon content of Europe generated by a generalized additive model. *European Journal of Soil Science*. doi:10.1111/ejss.12193.

Dixon, H., Hannaford, J. & Fry, M. J. 2013 The effective management of national hydrometric data: experiences from the United Kingdom. *Hydrological Sciences Journal* **58** (7), 1383–1399. doi:10.1080/02626667.2013.787486.

Duan, Q., Sorooshian, S. & Gupta, V. K. 1994 Optimal use of the SCE-UA global optimization method for calibrating watershed models. *Journal of Hydrology*. doi:10.1016/0022-1694(94)90057-4.

Environment Agency. 2012 *Review of Flow Derivation Method Denver Complex: Internal Report*.

Environment Agency 2013a *Cam and Ely Ouse Abstraction Licencing Strategy*. Available from: https://assets.publishing.service.gov.uk/government/uploads/system/uploads/attachment_data/file/637563/ALS_2017_Cam_and_Ely_Ouse.pdf.

Environment Agency 2013b *Witham Catchment Abstraction Management Strategy*. Available from: https://assets.publishing.service.gov.uk/government/uploads/system/uploads/attachment_data/file/289830/LIT7776_be21df.pdf.

Environment Agency 2017 *Witham at North Hykeham and Ely Ouse at Denver Flow Data*.

Environment Agency and National Resources Wales 2013 *Water Stressed Areas – Final Classification*. Available from: https://assets.publishing.service.gov.uk/government/uploads/system/uploads/attachment_data/file/244333/water-stressed-classification-2013.pdf.

European Environment Agency 2007 *CLC2006 Technical Guidelines, EEA Technical Report*. doi:10.2800/12134.

Francés, F., Vélez, J. I. & Vélez, J. J. 2007 Split-parameter structure for the automatic calibration of distributed hydrological models. *Journal of Hydrology*. doi:10.1016/j.jhydrol.2006.06.032.

Freer, J., Beven, K. & Ambroise, B. 1996 Bayesian estimation of uncertainty in runoff prediction and the value of data: an application of the GLUE approach. *Water Resources Research*. doi:10.1029/95WR03723.

Götzinger, J. & Bárdossy, A. 2007 Comparison of four regionalisation methods for a distributed hydrological model. *Journal of Hydrology* **333** (2–4), 374–384. doi:10.1016/j.jhydrol.2006.09.008.

Hannah, D. M., Demuth, S., van Lanen, H. A. J., Looser, U., Prudhomme, C., Rees, G., Stahl, K. & Tallaksen, L. M. 2011 Large-scale river flow archives: importance, current status and future needs. *Hydrological Processes* **25** (7), 1191–1200. doi:10.1002/hyp.7794.

He, Y., Bárdossy, A. & Zehe, E. 2011 A review of regionalisation for continuous streamflow simulation. *Hydrology and Earth System Sciences* **15** (11), 3539–3553. doi:10.5194/hess-15-3539-2011.

Hodge, I. & McNally, S. 2000 Wetland restoration, collective action and the role of water management institutions. *Ecological Economics*. doi:10.1016/S0921-8009(00)00171-3.

Hollis, J. M., Lilly, A., Higgins, A., Jones, R. J. A., Keay, C. A. & Bellamy, P. 2015 Predicting the water retention characteristics of UK mineral soils. *European Journal of Soil Science*. doi:10.1111/ejss.12186.

Holmes, M. G. R., Young, A. R., Gustard, A. & Grew, R. 2002 A region of influence approach to predicting flow duration curves within ungauged catchments. *Hydrology and Earth System Sciences* **6** (4), 721–731. doi:10.5194/hess-6-721-2002.

Hough, M. N. & Jones, R. J. A. 1997 The United Kingdom Meteorological Office rainfall and evaporation calculation system: MORECS version 2.0-an overview. *Hydrology and Earth System Sciences*. doi:10.5194/hess-1-227-1997.

Hrachowitz, M., Savenije, H. H. G., Blöschl, G., McDonnell, J. J., Sivapalan, M., Pomeroy, J. W., Arheimer, B., Blume, T., Clark, M. P., Ehret, U., Fenicia, F., Freer, J. E., Gelfan, A., Gupta, H. V., Hughes, D. A., Hut, R. W., Montanari, A., Pande, S., Tetzlaff, D., Troch, P. A., Uhlenbrook, S., Wagener, T., Winsemius, H. C., Woods, R. A., Zehe, E. & Cudennec, C. 2013 A decade of predictions in ungauged basins (PUB) – a review. *Hydrological Sciences Journal*. doi:10.1080/02626667.2013.803183.

Ireson, A. M., Mathias, S. A., Wheater, H. S., Butler, A. P. & Finch, J. 2009 A model for flow in the chalk unsaturated zone incorporating progressive weathering. *Journal of Hydrology*. doi:10.1016/j.jhydrol.2008.11.043.

Kiang, J. E., Gazoorian, C., McMillan, H., Coxon, G., Le Coz, J., Westerberg, I. K., Belleville, A., Sevrez, D., Sikorska, A. E., Petersen-Øverleir, A., Reitan, T., Freer, J., Renard, B., Mansanarez, V. & Mason, R. 2018 A comparison of methods for streamflow uncertainty estimation. *Water Resources Research*. doi:10.1029/2018WR022708.

Lanfear, K. J. & Hirsch, R. M. 1999 USGS study reveals a decline in long-record streamgages. *Eos*. doi:10.1029/99EO00406.

Le Coz, J., Renard, B., Bonnifait, L., Branger, F. & Le Boursicaud, R. 2014 Combining hydraulic knowledge and uncertain gaugings in the estimation of hydrometric rating curves: a Bayesian approach. *Journal of Hydrology*. doi:10.1016/j.jhydrol.2013.11.016.

Lewis, M., Cheney, C. S. & O Dochartaigh, B. E. 2006 Guide to permeability indices. *British Geological Survey*.

McMillan, H., Freer, J., Pappenberger, F., Krueger, T. & Clark, M. 2010 Impacts of uncertain river flow data on rainfall-runoff model calibration and discharge predictions. *Hydrological Processes* **24** (10), 1270–1284. doi:10.1002/hyp.7587.

McMillan, H., Krueger, T. & Freer, J. 2012 Benchmarking observational uncertainties for hydrology: rainfall, river discharge and water quality. *Hydrological Processes* **26** (26), 4078–4111. doi:10.1002/hyp.9384.

Mlynowski, T. J., Hernández-Henríquez, M. A. & Déry, S. J. 2011 An evaluation of hydrometric monitoring across the Canadian pan-Arctic region, 1950–2008. *Hydrology Research* **46** (6), 479. doi:10.2166/nh.2011.105.

Mott MacDonald 2007 *Development of Rainfall-Runoff Models for the Nar, Wissey and Wensum Catchments.*

Mott MacDonald 2017 *Black Sluice IDB Water Balance Assessment Methodology.*

Moulin, L., Gaume, E. & Obled, C. 2009 Uncertainties on mean areal precipitation: assessment and impact on streamflow simulations. *Hydrology and Earth System Sciences.* doi:10.5194/hess-13-99-2009.

Nalbantis, I., Efstratiadis, A., Rozos, E., Kopsiafti, M. & Koutsoyiannis, D. 2011 Holistic versus monomeric strategies for hydrological modelling of human-modified hydrosystems. *Hydrology and Earth System Sciences.* doi:10.5194/hess-15-743-2011.

Nash, J. E. & Sutcliffe, J. V. 1970 River flow forecasting through conceptual models part I – a discussion of principles. *Journal of Hydrology.* doi:10.1016/0022-1694(70)90255-6.

National River Flow Archive 2017 *NRFA Gauging Station Data.* Available from: https://nrfa.ceh.ac.uk/data (accessed 16 August 2017).

National Rivers Authority. 1993 *Ely Ouse Catchment Management Plan.*

Neff, E. L. 1977 How much rain does a rain gage gage? *Journal of Hydrology.* doi:10.1016/0022-1694(77)90001-4.

Oudin, L., Perrin, C., Mathevet, T., Andréassian, V. & Michel, C. 2006 Impact of biased and randomly corrupted inputs on the efficiency and the parameters of watershed models. *Journal of Hydrology.* doi:10.1016/j.jhydrol.2005.07.016.

Reed, S., Koren, V., Smith, M., Zhang, Z., Moreda, F., Seo, D.-J. & DMIP Participants, A. 2004 Overall distributed model intercomparison project results. *Journal of Hydrology* **298** (1–4), 27–60. doi:10.1016/j.jhydrol.2004.03.031.

Refsgaard, J. C. 1997 Parameterisation, calibration and validation of distributed hydrological models. *Journal of Hydrology.* doi:10.1016/S0022-1694(96)03329-X.

Renard, B., Kavetski, D., Leblois, E., Thyer, M., Kuczera, G. & Franks, S. W. 2011 Toward a reliable decomposition of predictive uncertainty in hydrological modeling: characterizing rainfall errors using conditional simulation. *Water Resources Research.* doi:10.1029/2011WR010643.

Rodda, J. C. 1998 Hydrological networks need improving! In: *Water: A Looming Crisis? International Conference on World Water Resources at the Beginning of the 21st Century* (H. Zebidi, ed.). UNESCO, Paris, pp. 91–102.

Saxton, K. E. & Rawls, W. J. 2006 Soil water characteristic estimates by texture and organic matter for hydrologic solutions. *Soil Science Society of America Journal* **70** (5), 1569–1578.

Sene, K. J. & Farquharson, F. A. K. 1998 Sampling errors for water resources design: the need for improved hydrometry in developing countries. *Water Resources Management* **12** (2), 121–138. doi:10.1023/A:1007952319495.

Silberstein, R. P. 2006 Hydrological models are so good, do we still need data? *Environmental Modelling and Software* **21** (9), 1340–1352. doi:10.1016/j.envsoft.2005.04.019.

Sivapalan, M., Takeuchi, K., Franks, S. W., Gupta, V. K., Karambiri, H., Lakshmi, V., Liang, X., McDonnell, J. J., Mendiondo, E. M., O'Connell, P. E., Oki, T., Pomeroy, J. W., Schertzer, D., Uhlenbrook, S. & Zehe, E. 2003 IAHS decade on predictions in ungauged basins (PUB), 2003–2012: shaping an exciting future for the hydrological sciences. *Hydrological Sciences Journal* **48** (6), 857–880. doi:10.1623/hysj.48.6.857.51421.

UPV-GIMHA 2014 *Description of the Distributed Conceptual Hydrological Model Tetis V.8.* Universitat Politècnica de València, Valencia.

Vélez, J. J., Puricelli, M., López Unzu, F. & Francés, F. 2009 Parameter extrapolation to ungauged basins with a hydrological distributed model in a regional framework. *Hydrology and Earth System Sciences.* doi:10.5194/hess-13-229-2009.

von Lany, P. H., Choudhury, F., Hepworth, N. & Akande, K. 2013 Applying optimisation and uncertainty analysis to help develop an integrated water resources plan for South East England. *Water Resources Management.* doi:10.1007/s11269-012-0121-2.

Vrugt, J. A., Gupta, H. V., Bouten, W. & Sorooshian, S. 2003 A shuffled complex evolution metropolis algorithm for optimization and uncertainty assessment of hydrologic model parameters. *Water Resources Research.* doi:10.1029/2002WR001642.

Vrugt, J. A., ter Braak, C. J. F., Gupta, H. V. & Robinson, B. A. 2009 Equifinality of formal (DREAM) and informal (GLUE) Bayesian approaches in hydrologic modeling? *Stochastic Environmental Research and Risk Assessment.* doi:10.1007/s00477-008-0274-y.

Wagener, T. & Montanari, A. 2011 Convergence of approaches toward reducing uncertainty in predictions in ungauged basins. *Water Resources Research.* doi:10.1029/2010WR009469.

Wagener, T., Lees, M. J. & Wheater, H. S. 2002 A toolkit for the development and application of parsimonious hydrological models. *Water Resources Publications* **2**, 34.

Wagener, T., Wheater, H. S. & Gupta, H. V. 2004 *Rainfall-Runoff Modelling in Gauged and Ungauged Catchments.* Imperial College Press, London.

Wilby, R. L. 2005 Uncertainty in water resource model parameters used for climate change impact assessment. *Hydrological Processes.* doi:10.1002/hyp.5819.

Young, A. R., Grew, R. & Holmes, M. G. R. 2003 Low flows 2000: a national water resources assessment and decision support tool. *Water Science and Technology* **48** (10), 119–26.

First received 17 January 2019; accepted in revised form 14 May 2019. Available online 3 June 2019

Integration of hillslope hydrology and 2D hydraulic modelling for natural flood management

Barry Hankin, Peter Metcalfe, Keith Beven and Nick A. Chappell

ABSTRACT

Natural flood management (NFM) has recently invigorated the hydrological community into redeploying its process understanding of hydrology and hydraulics to try to quantify the impacts of many distributed, 'nature-based' measures on the whole-catchment response. Advances in spatial data analysis, distributed hydrological modelling and fast numerical flow equation solvers mean that whole-catchment modelling including computationally intensive uncertainty analyses are now possible, although perhaps the community has not yet converged on the best overall parsimonious framework. To model the effects of tree-planting, we need to understand changes to wet canopy evaporation, surface roughness and infiltration rates; to model inline storage created by 'leaky barriers' or offline storage, we need accurate channel hydraulics to understand the changes to attenuation; to model the complex behaviour of the whole network of NFM measures, and the possibility of flood peak synchronisation effects, we need efficient realistic routing models, linked to key flow pathways that take into account the main physical processes in soils and the antecedent moisture conditions for a range of different rainfall events. This paper presents a new framework to achieve this, based on a cascade of the Dynamic Topmodel runoff generation model and the JFlow or HEC-RAS 2D hydraulic models, with an application to the Swindale Catchment in Cumbria, UK. We demonstrate the approach to quantify both the effectiveness of a relatively large 'runoff attenuation feature' in the landscape and the uncertainty in the calculation given model parameter uncertainty.

Key words | hydraulics, hydrology, NFM, Topmodel, uncertainty

Barry Hankin (corresponding author)
Peter Metcalfe
Keith Beven
Nick A. Chappell
Visiting Researcher, Lancaster Environment Centre,
Lancaster University,
Bailrigg, Lancaster LA14YW,
UK
E-mail: *barry.hankin@jbaconsulting.com*

INTRODUCTION

There has been a growing number of investigations attempting to model the effectiveness of NFM measures for flood risk reduction (Nicholson *et al.* 2012; Ghimire *et al.* 2015; Metcalfe *et al.* 2017, 2018), although relatively few have focussed on modelling strategy (Hankin *et al.* 2017a). Such measures comprise distributed changes in the landscape aiming to emulate natural processes to store, slow or infiltrate water to help reduce flooding and have been collectively referred to as working with natural processes (Pitt 2008; Burgess-Gamble *et al.* 2017), natural flood management (UK and used hereinafter), Natural Water Retention Measures (Europe) and Nature-Based Approaches (international, e.g. World Wildlife Fund 2016). Such measures include in-stream leaky barriers (Metcalfe *et al.* 2017), engineered log jams (Addy & Wilkinson 2016), riparian or widespread tree-planting (Carrick *et al.* 2018), runoff attenuation features (RAFs) in the landscape (Nicholson *et al.* 2012; Metcalfe *et al.* 2018), peatland restoration to increase roughness (Holden *et al.* 2008; Shuttleworth *et al.* 2019), and river restoration including re-meandering and floodplain

doi: 10.2166/nh.2019.150

reconnection to hold back and temporarily store more water which has been popular across Europe (e.g. Office International de l'Eau 2013). They present a challenge to hydrologists and hydraulic modellers to first identify the catchment processes that might be altered, and then to justify changes or 'shifts' to effective parameter values controlling these processes in their models.

NFM has often been seen as a low-cost alternative to traditional flood risk management techniques and is being deployed by communities at risk unable to secure funding for measures such as hard flood defences. Many such examples are included within 65 UK case studies recently compiled in Burgess-Gamble *et al.* (2017), with online mapping (naturalprocesses.jbahosting.com/) and a dedicated European website (http://nwrm.eu/). The recent UK Evidence Directory (Burgess-Gamble *et al.* 2017) and review papers (Dadson *et al.* 2017; Lane 2017) summarise the current state of knowledge, pointing to evidence at the small (10 km^2) to medium (100 km^2) scales and for small (<10% annual exceedance probability – AEP) to medium floods (1% < AEP < 10%), but indicate little direct evidence for effectiveness at larger scales (>100 km^2) and major flood events (<1% AEP). European research (Office International de l'Eau 2013) and other international advice on nature-based methods (World Wildlife Fund 2016) provide much advice on incorporating risk reduction measures based on green engineering.

Strategies for distributed modelling of NFM (e.g. Hankin *et al.* 2017a) and testing the resilience against realistic spatial extremes (Hankin *et al.* 2017b) and synchronisation effects (Pattison *et al.* 2014) rely on modellers first being able to represent distributed hydrological responses accurately, and then represent the changes incurred to those processes as a result of small-scale distributed nature-based interventions realistically. Since the effective parameterisation of these processes and changes to processes are very uncertain, a framework for uncertainty analysis such as the generalised likelihood uncertainty estimation methodology (Beven & Binley 1992; Beven & Binley 2014) becomes a necessity, and methods for detecting change in ensemble modelling results with and without nature-based measures important (Hankin *et al.* 2017a).

Physics-based models can offer a useful lens for upscaling small-scale distributed processes such as increases in friction, storage and infiltration to study their combined effect at the catchment scale, yet also reveal a blurry image of our knowledge of exactly how to represent often variable and difficult-to-measure processes. The crux for modellers has therefore been how to represent very distributed and evidenced small-scale changes to hydraulic and hydrological processes, using efficient tools to capture the main interactions, whilst allowing for ensemble simulations and uncertainty estimation. Often a single type of model is used to represent the whole system, whereas using a combination of models (which are becoming more versatile to integrate) can improve the overall accuracy of process representation. The 'evidenced' changes to effective distributed parameters require further research, currently the focus of three ongoing UK Natural Environmental Research Council research projects (https://nerc.ukri.org/research/funded/programmes/nfm/).

This paper demonstrates a new integrated modelling framework, isolating one large runoff attenuation feature in a relatively small catchment, to test the effectiveness of a new modelling strategy to assess changes in catchment response whilst also quantifying uncertainty. The paper aims to assert the need for such a framework with the following three objectives:

- Bring together two fast technologies that improve integrated modelling of the distributed hydrology and hydrodynamics required to simulate small-scale NFM interventions and scale the effects up to the catchment scale, whilst allowing for ensemble simulations to explore uncertainties. Hillslope processes are modelled using the R version of Dynamic Topmodel (Metcalfe *et al.* 2016) with channel and floodplain hydrodynamics modelled using JFlow (Lamb *et al.* 2009).
- Undertake cascade modelling of an ensemble of 'acceptable' model structures (Beven 2006; Blazkova & Beven 2009) and define the modelling uncertainties in our estimates of the effectiveness of NFM. Many investigations are using modelling to scale up the effects of distributed catchment measures, to understand the change in response at the outlet, but few are attempting to *quantify uncertainties in their answers*.
- Use more of the 2D modelling outputs to drive a better understanding of processes having a strong influence on flooding and flood risk reduction – here we use 2D

depth and velocity data to estimate shear stresses and generate erodibility maps.

Study area and data

Swindale is a small, 15 km^2 upland valley near Shap in Cumbria, UK, with water draining from upland peatland cascading down to a valley with runoff characteristics dominated by impermeable glacial till. It has a high annual average rainfall (standard period annual average rainfall (SAAR) = 2,445 mm) and low baseflow index (BFIHOST = 0.29). Flows are gauged upstream of the United Utilities (UU) abstraction point (NY5146113169) that feeds water to Haweswater in the adjacent valley. The UU flow gauge (station 761113) has telemetry at 15 min 1997–2015; 2016 to present, with a gap whilst the intake and fish pass were upgraded. This unfortunately results in no data for the extreme event of Storm Desmond in winter 2015, so a large storm November 2009 is studied here with an estimated annual exceedance probability of between 50% and 20% of the scale where there is more evidence for NFM interventions being effective (Burgess-Gamble *et al.* 2017).

The Swindale catchment has been the subject of recent investigations by UU and the Royal Society for the Protection of Birds (RSPB) into land management and peatland restoration strategies, river restoration and riparian and upland tree-planting. The effects of these distributed NFM measures were previously modelled (RSPB 2017), identifying that the most effective single measure was a large RAF created by restricting flow out of a large storage area at Dodd Bottom (marked as blue in Figure 1). For the purposes of modelling at a fine temporal resolution, the closest 15-minute Tipping Bucket Rainfall gauge in the vicinity was Deeming Moss (NGR NY5545206553, SAAR = 1,982 mm), which was proportionately scaled using the annual average rainfall based on catchment descriptors, although there will be unknown local orographic and rain shadow effects. A digital terrain model (DTM) with detailed resolution (0.5 m) was available from the RSPB (collected by Glasgow University), in the vicinity of Dodd Bottom and the restored river meandering. This was resampled to 2 m resolution using bilinear interpolation, and feathered into coarser scale topography (2 m composite DTM based on LiDAR and where unavailable, Synthetic Aperture Radar) to increase the computational efficiency of the model calculations.

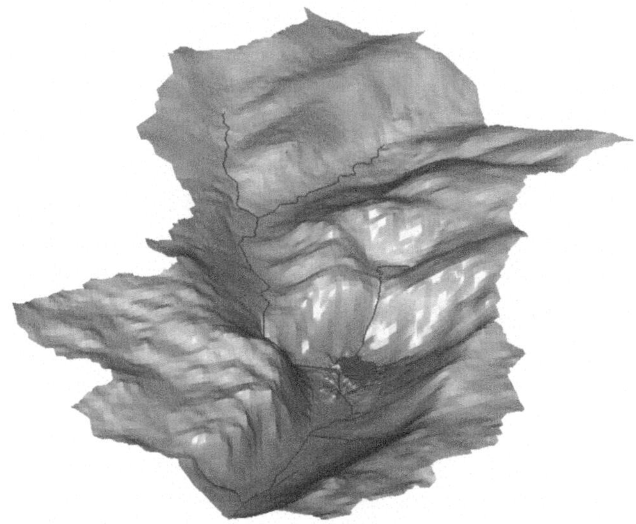

Figure 1 | Three-dimensional schematic of Swindale topography with greater detail in the valley bottom, and where the Dodd Bottom storage area is marked in blue context in Figure 2. Please refer to the online version of this paper to see this figure in colour: http://dx.doi.org/10.2166/nh.2019.150.

METHOD

We demonstrate a cascade modelling approach that couples Dynamic Topmodel (Beven & Freer 2001; Metcalfe *et al.* 2016) with JFlow (Lamb *et al.* 2009). It takes advantage of the computational efficiency and parsimonious parameterisation of hillslope hydrology in Dynamic Topmodel, coupled with the fast 2D hydrodynamic modelling offered by the graphics-card implementation of the Shallow Water Equations (SWE) in JFlow (for benchmarking study, see Hunter *et al.* 2008 and the updated speed testing for the SWE solver in Lamb *et al.* 2009). This highly efficient combination allows us to undertake ensemble simulations to explore uncertainties, whilst capturing essential surface/subsurface interactions in the wider catchment, *and* the complex channel–floodplain hydrodynamic interactions needed to understand the influence of in-channel NFM measures. Other rainfall runoff models could be used in a similar approach, but here Dynamic Topmodel has been used based on it having a low number of parameters, whilst being able to represent the key processes in a set of distributed hydrological response units (HRUs) which dynamically interact with one another. Many rainfall runoff models such as PDMs, hydrological predictions for the environment (HYPE) and SWAT similarly use efficient HRUs to generate responses reflective of local topography or soils, but few use a flow distribution matrix to pass flows between the landscape units to more properly reflect celerities in the saturated zone responses. In the R version of Dynamic Topmodel (Metcalfe *et al.* 2016), saturation excess overland flow can be generated in an upslope HRU and be passed downslope to an HRU with a deficit and re-infiltrate. Here, we require a general NFM modelling framework that can represent such processes since these are processes that we seek to influence through, for instance, tree-planting or reduced grazing intensity. In the discussion we also demonstrate a similar coupling between Dynamic Topmodel and HEC-RAS2D, giving similar results and highlighting the increasing versatility of 2D hydrodynamic packages.

The model cascade is a natural extension of the modelling in the Brompton catchment (Metcalfe *et al.* 2017), in which the distributed runoff from Dynamic Topmodel was fed into a network solution of the 1D St Venant equations to explore leaky barriers, that were designed largely to store water in the channel. Here, we require that the general modelling framework is capable of representing the effects of NFM interventions that promote additional flood flow attenuation by encouraging water onto the floodplain where it was previously poorly connected. The method also builds upon the direct rainfall with losses 2D modelling approach (e.g. Hankin *et al.* 2017a) that use Revitalised Flood Hydrograph losses model (Kjeldsen *et al.* 2005) or approaches that simulate distributed infiltration using an effective hydraulic conductivity. None of these approaches are capable of modelling surface–subsurface interactions, such as saturation excess runoff flowing overland until it reaches an unsaturated area, whereupon it re-infiltrates (run-on). Such interactions can, however, be represented in Dynamic Topmodel, so here we combine this more realistic hillslope process representation with the 2D hydrodynamic routing to capture the complexity of flows over steep and diverse terrain and distributed interactions between channel and floodplain, where many NFM measures are currently being experimented with (for examples, see Office International de l'Eau 2013; Clilverd *et al.* 2013). The cascade of models, therefore, provides a useful compromise between detail in process representations and computational efficiency and a framework for exploring NFM and uncertainties.

Before Dynamic Topmodel is used to simulate a catchment, some preprocessing of the Digital Terrain Model (Figure 1) is required in order to divide the catchment into areas that respond in a hydrologically similar way. These are the HRUs (shown in Figure 2) in Dynamic Topmodel and are typically based on accumulated upslope area, slope or distance to a channel. A flow connectivity matrix provides the relationship between the direct runoff between these HRUs and into the channel, represented initially through a time delay histogram in order to route the flows to the outlet, and permitting an assessment of performance at the UU flow gauge. Driven by rainfall and evapotranspiration for the winter storm modelled (November 2009), the different parameters controlling the runoff response in the HRUs were sampled from uniform random distributions between typical ranges expected for this upland catchment. Swindale has peat in the uplands and slowly permeable, shallow soils lower down dominated by glacial till deposits synonymous with slowly permeable soils. More details of the parameters in the Dynamic Topmodel model are provided in Beven & Freer (2001), but the most sensitive parameters include the saturated surface transmissivity, the exponential transmissivity shape parameter, m, the route zone storage coefficient and the vertical time delay parameter, Td.

Solutions meeting a relatively high Nash–Sutcliffe efficiency (NSE) (>0.85) were classed as behavioural. The more stringent limits of acceptability approach (Beven 2006; Blazkova & Beven 2009; Liu *et al.* 2009; Hollaway *et al.* 2018) was not used at this stage, but the intention is to use this for new applications to larger catchments. Insisting on a relatively high performance ($NSE > 0.85$) resulted in 1,482 behavioural simulations out of the ten thousand Monte-Carlo simulations that were undertaken, and these were taken forward and fed into the 2 m resolution JFlow model. This was achieved by taking the simulated distributed runoff (subsurface and surface flows) and feeding the time series into along-reach internal flow boundaries in the 2 m resolution JFlow mesh, represented in Figure 2. Here, the colour shading represents the different HRUs where each produce similar surface and subsurface runoff responses based on the model of Metcalfe *et al.* (2016). In this investigation, eight in-stream reaches were used, represented as HRUs that receive flows via the flow distribution matrix in Dynamic Topmodel. These time series are distributed over internal flow boundaries as through-time source terms in the SWE in JFlow.

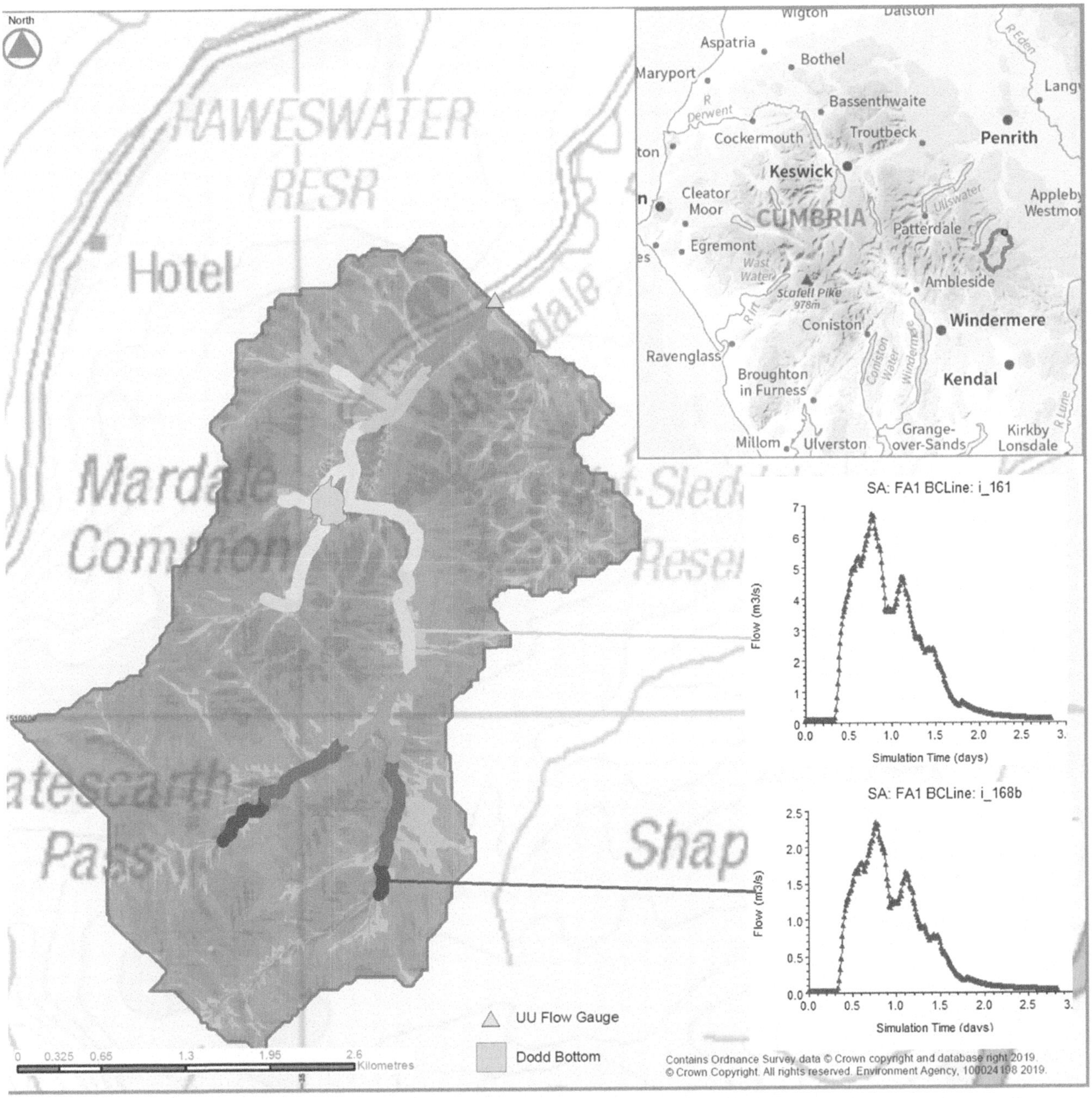

Figure 2 | Schematic of cascade modelling approach showing shading based on hillslope and channel-based hydrological response units and internal flow boundaries for the hydrodynamic model with example lateral inflows from Dynamic Topmodel. Please refer to the online version of this paper to see this figure in colour: http://dx.doi.org/10.2166/nh.2019.150.

In the discussion, we also compare the same approach to model integration using a combination of Dynamic Topmodel and HEC-RAS 2D (using a diffusion wave), using a much coarser resolution numeric grid (10 m). We discuss how this scheme is able to incorporate detailed subgrid topographic data (2 m) despite the coarse grid (10 m), and explore additional outputs from the 2D hydrodynamic models, such as generating spatial distributions of shear stresses and erodibility maps. These outputs can help understand sediment processes that are also important to understanding risk reduction and the long-term performance of RAF-based NFM measures, which may, if designed incorrectly fill up with sediment.

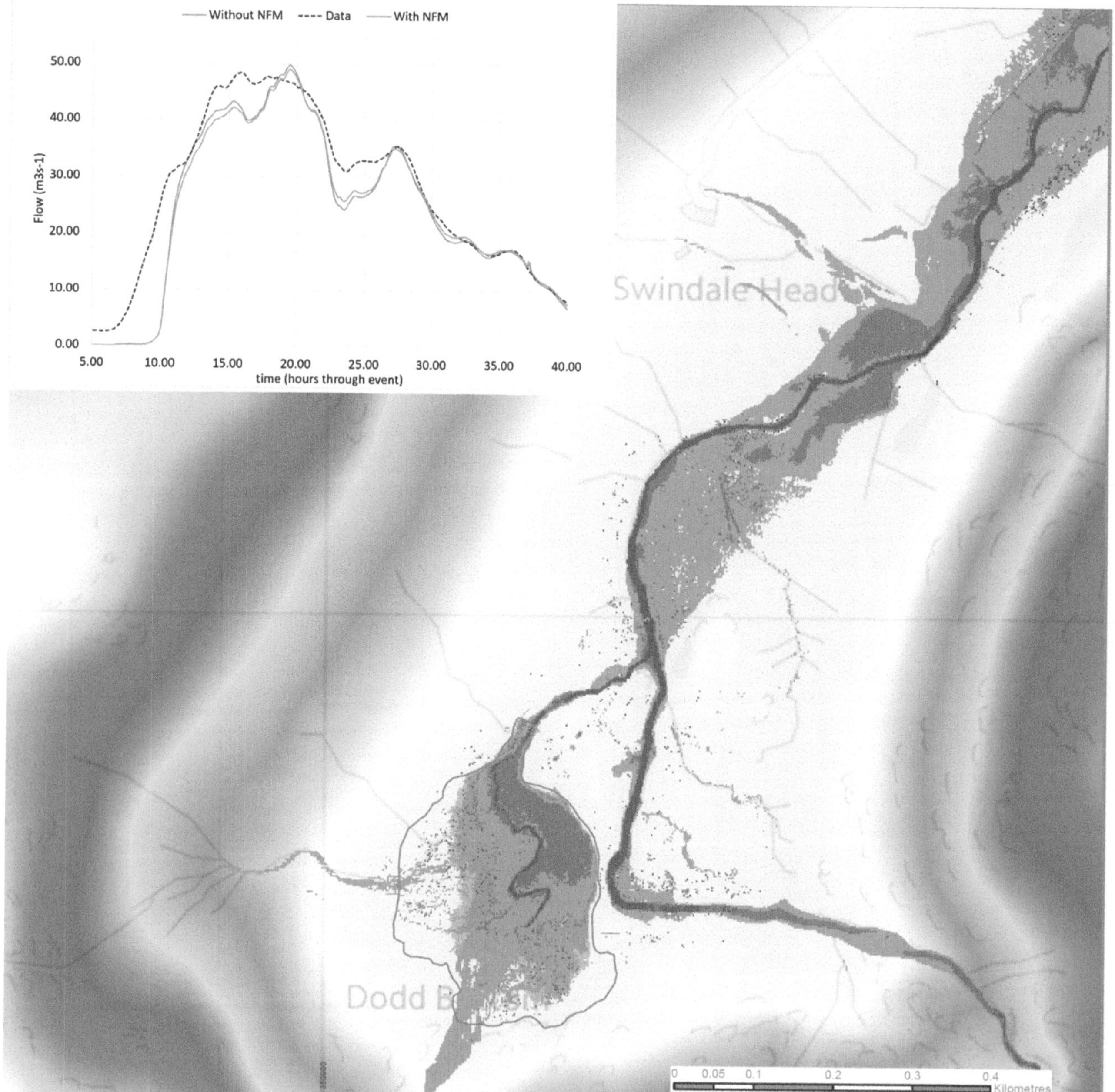

Figure 3 | Cascade modelling for the baseline modelling (blue) with a simulated NFM intervention (red) depth grids and hydrograph showing performance at UU flow gauge inset. Volume change 16,100 m^3 to 51,700 m^3 with NFM following the intervention. Please refer to the online version of this paper to see this figure in colour: http://dx.doi.org/10.2166/nh.2019.150.

RESULTS

Figure 3 shows the initial spatiotemporal results of a single coupled simulation, with and without an in-channel obstruction downstream of Dodd Bottom (Figure 1) to throttle flows using a reduced outflow width of 4 m so as to store more peak flows temporarily in the areas known as Dodd Bottom labelled in the lower left. We focus effort here on this single large RAF and model it in isolation from other measures to test the framework. The re-meandering of the watercourse in the valley bottom undertaken to improve biodiversity is included in both scenarios and can be seen in darker blue (deeper) water in Figure 3 where it is evident that the floodplain in this area is inundated for the

November 2009 event. Critically, there is no evidence of unrealistic stepping in the predicted depths near the boundaries between the river reaches receiving different time series from Dynamic Topmodel, apart from close to the intervention, where the red shaded depth grid shows the additional storage due to constricting the width of the outlet to the RAF to 4 m. Use of Dynamic Topmodel alone would not have been able to accurately represent and show this level of important detail in the floodplain. The inset hydrograph shows a slight underprediction compared with the observed 48.3 $m^3\,s^{-1}$ peak flow at the UU river gauging station, although the ensemble results are analysed further below, showing a range of under- and overpredictions of the peak.

A sample of 120 of the 1,482 ensemble baseline behavioural simulations is shown in Figure 4 highlighting a range of peak flows in comparison with the gauged flow. The overall performance is strong (NSE ~ 0.8–0.9), although there is a delay in the early part of the rising limb for all the model runs and a slight difference in response near the peak with observations suggesting a broader peak.

Rather than plot the time series for the intervention scenario, it is more informative to look at the difference between baseline and intervention (following Metcalfe *et al.* 2018). Figure 5 shows the 'difference hydrograph' between the baseline and the RAF intervention scenario for the sample of 120 simulations, for which positive values represent a reduction on the rising limb of the hydrograph, and negative values indicate successfully attenuated flows returning back to the watercourse. The maximum difference in the peak is 1.8 $m^3\,s^{-1} \pm 1.4\,m^3\,s^{-1}$, although the largest changes tend to occur in advance of the peak in the hydrograph that occurs around hour 20 at the UU flow gauge. This confirms what was suspected from observations of the basic geometry of the watercourse (see Figure 1), in that Dodd Bottom is on a faster-rising part of the catchment, and the additional attenuation tends to result in taking volumes from the rising limb of the hydrograph as considered at the UU flow gauge.

The reduction in the peak flow afforded by the additional temporary storage is not so pronounced by the time the flows reach the UU flow gauge (Figure 5). This indicates that the restriction applied (4 m) needs to be narrower, and that restricting flows in this area from what is a faster rising part of the catchment is not an effective strategy and to improve it further, flows from the main channel may also need to be diverted into Dodd Bottom and multiple interventions may be needed. However, this is a sensitive

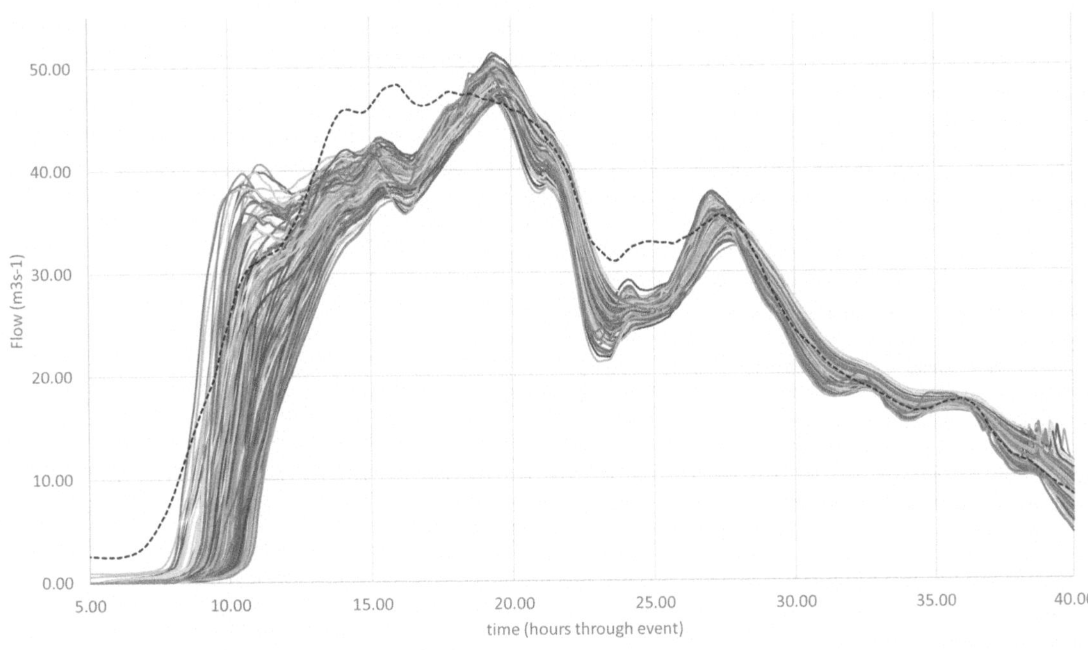

Figure 4 | Sample of 120 behavioural simulations for the baseline scenario at flow gauge. Observed flows in black-dashed. Simulations are randomly coloured to highlight variation within the ensemble of acceptable models.

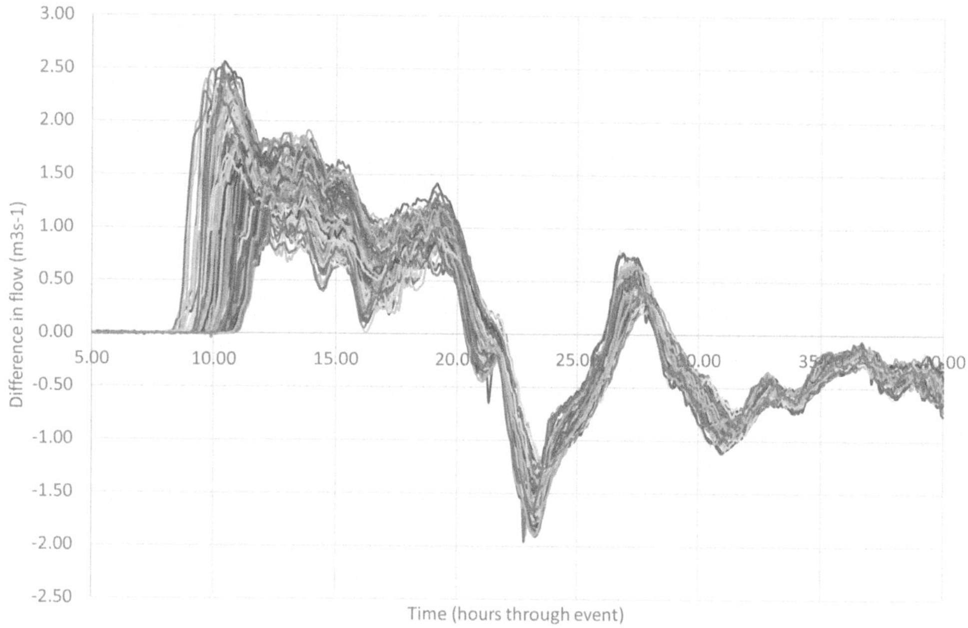

Figure 5 | Difference hydrograph: Baseline hydrograph minus the intervention hydrograph. Peak difference = 1.8 m^3 s^{-1} ± 1.4 m^3 s^{-1}. simulations randomly coloured to highlight variation in the ensemble of acceptable models. Please refer to the online version of this paper to see this figure in colour: http://dx.doi.org/10.2166/nh.2019.150.

site, and building bunds, structures and diversions across unique moraine formations is not likely to be acceptable, with the added constraint that the volume retained if impounded would come under the UK Reservoirs Act and thereby require regulation.

In addition to the spaghetti time series plots of the ensemble predictions in Figure 5, it is important to view the ensemble results spatially and develop methods for constraining model outputs which can also be helpful for expressing additional limits on the acceptability of model structures. An example of this can be found in Aronica *et al.* (1998) where the approximate timing of part of the flood wave exiting one area of the Imera flood basin was used as a fuzzy acceptability criteria, which if not simulated, was used to effectively reject models that might otherwise have generated an acceptable behaviour.

To understand the differences between Monte-Carlo simulations, it is once again more informative to look at the differences in predicted depths in this catchment, since the high performing simulations (NSE > 0.85) have similar extents. Figure 6 shows the performance-weighted depth grid in the central pane for the NFM scenario (6e–on scale of 0.0–3.0 m), with the eight outside panes showing the difference between this most likely depth grid and eight other NFM Monte-Carlo simulations (panes 6a–6i *excluding* 6e on a scale of −0.1 to +0.1 m). Some of these show positive differences (6a, 6 h, 6i), negative differences (6c, 6d, 6 g), and some not much difference (6b, 6f), but the range of predicted depths is expected based on the range of differences in Figure 5. This example illustrates how there can be differences in spatial behaviour, including potentially flood pathways within our acceptable simulations, which may coincide with knowledge of flooding that can be used to constrain the model behaviour further and reject certain modes.

DISCUSSION

The key advantages of the new approach demonstrated here include efficient, distributed representation of hillslope processes using Dynamic Topmodel, combined with more accurate 2D hydraulics of flows over a complex terrain. The previous coupling of Dynamic Topmodel with 1D hydraulic routing has proven successful in a similar scale UK catchment (Metcalfe *et al.* 2017), including representation of the hydraulics of in-stream leaky barriers. Such measures are also being designed to spill flows onto the

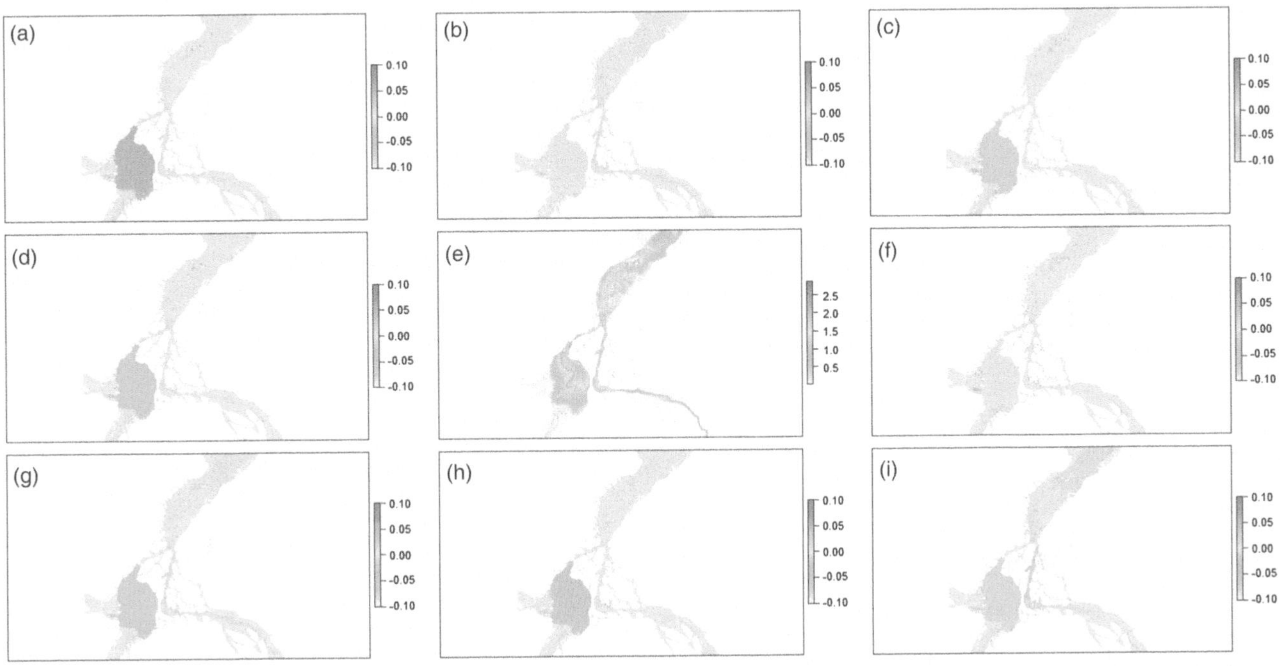

Figure 6 | Performance (NSE)-weighted depth grid (6e on a scale of 0.0 m to 3.0 m), surrounded by plots showing the depth differences (−0.1 m to +0.1 m) between weighted grid and eight other simulations.

floodplain to enhance flood attenuation, and so a 2D model becomes more desirable. The use of the fast 2D SWE model, JFlow, allows the impacts of in-channel modifications such as 'leaky barriers' and floodplain reconnection, and failures of these features, to be more adequately explored at larger scales. This includes assessing their positive and negative effects on sediment transport, in ways that take into account the large uncertainties inherent in the modelling process (including the parameters representing change within the models).

To show that the coupling Dynamic Topmodel with JFlow is a generic approach, we present a preliminary simulation of Dynamic Topmodel with an alternative 2D hydraulic model, namely, the licence-free HEC-RAS2D (https://www.hec.usace.army.mil/software/hec-ras/downloads.aspx). The HEC-RAS2D model has a new type of numerical scheme that permits efficiencies, whereby a relatively coarse numerical grid can be used (e.g. 10 m), for which the subgrid topography (2 m) is taken into account. This is achieved through precomputing hydraulic tables for each coarse numerical grid cell, comprising the conveyance across each face as a function of subgrid geometry, and then also a table of storage volume in the cell as a function of water elevation. This permits the modelling of channels that pass through the numerical cells if they are represented in the subgrid topography. Figure 7 shows the results of using HEC-RAS2D with a 10 m numerical grid based on a 2 m resolution subgrid topography compared with JFlow. The solution shown is driven by 8 time series distributed along 8 inflow reaches that are identified as HRUs in the Dynamic Topmodel modelling scheme. The red-dashed HEC-RAS 2D output is compared with the gauged data (black dash) and one of the JFlow ensemble runs. The slight differences between JFlow and HEC-RAS2d are likely to stem from the different resolutions: JFlow uses a 2 m grid, HEC-RAS 2d uses a 10 m grid with hydraulic tables based on the 2 m sub-grid topography; and the equations: HEC-RAS 2D uses an implicit solution to the Diffusion Wave, whilst JFlow solves the full SWE using an explicit scheme.

The approach yields very similar behaviour and performance at the gauge, and this increases confidence in the outputs from both models that either could be used in this model cascade approach.

To further illustrate the advantages of using 2d hydraulic models to better represent channel/floodplain flows resulting from the hillslope outputs Dynamic Topmodel, we very

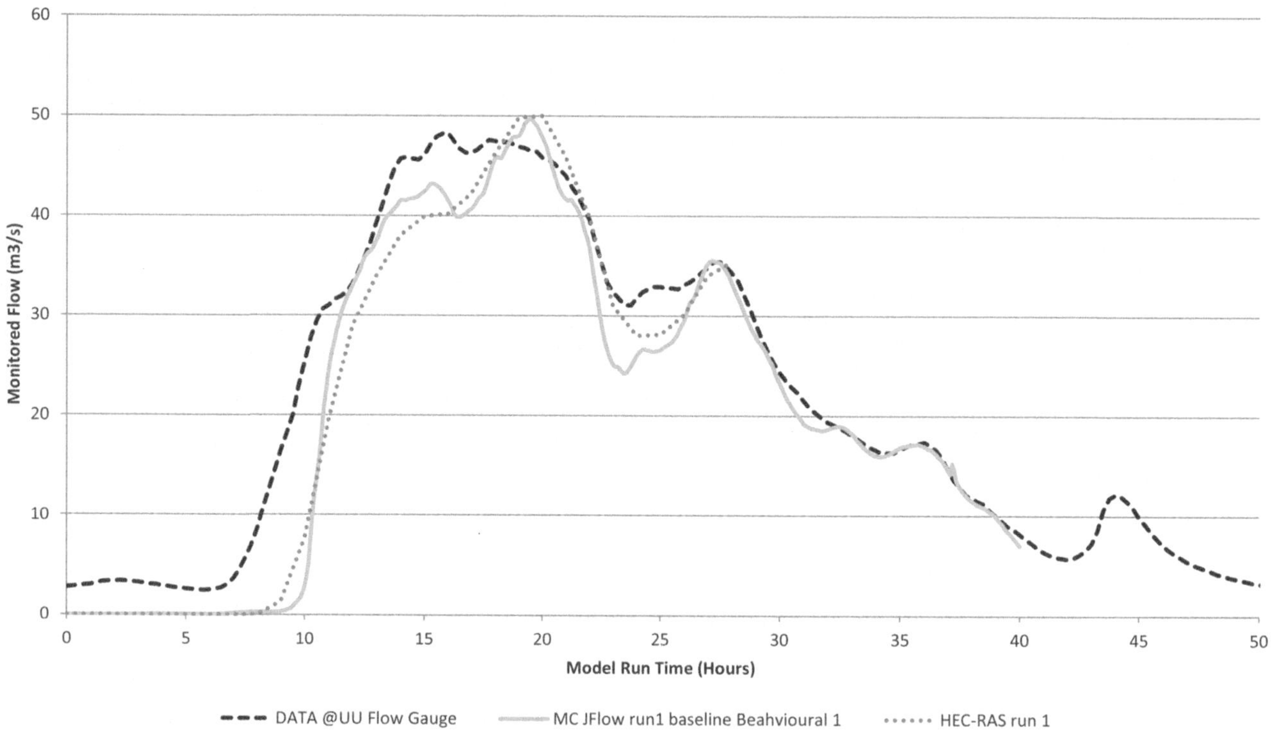

Figure 7 | Comparisons between two 2D hydraulic models for the baseline scenario. Please refer to the online version of this paper to see this figure in colour: http://dx.doi.org/10.2166/nh.2019.150.

briefly show the ease to which the hydraulic modelling could be extended to understand the spatial distribution of shear stresses and sediment erodibility, critical to understanding the performance of NFM measures longer term. A problem with RAF measures is the potential for reduction of flow velocities and subsequent sediment deposition within the temporary storage area, or localised increases in velocity at the outlet to the storage area, creating scour and unwanted bed erosion, which can both lead to a loss of performance in terms of flood risk reduction.

Given our framework uses 2D hydrodynamic modelling in areas of significant flow, this can be used to provide velocity and shear stress outputs that can help with the design of NFM to avoid in-filling. Here, we follow approaches discussed in Lane & Ferguson (2005), and more recently Reid *et al.* (2018), whereby we use the distributed depth and velocity information from JFlow or HEC-RAS2D, combined with Manning's roughness to estimate shear stress of the form:

$$\tau = \frac{\rho g n^2 U^2}{d^{1/3}} \quad (1)$$

where U is the depth-averaged velocity, d is the depth, and n is Manning's roughness. This is dimensionally consistent with shear stress and produces a similar shear stress compared with assuming a logarithmic vertical velocity profile. In a comparative exercise, the distributed value of shear stress was then compared with the critical shear stress for movement of bed material, given an estimate of average grain size. The RSPB provided estimates of $D_{50} = 82$ mm in upper Swindale. Using a Shields constant of 0.06, the zones of likely erosion are developed, shown in red in Figure 8. A zone of deposition was also derived using a deposition coefficient multiplier of 0.3 to the critical shear stress for the bedload and shown in yellow in Figure 8. Areas between these values are assumed to be a zone of transition (in blue). Of note, is a large red patch of a zone of likely erosion at the start of where the channel re-meandering occurs (flows are from bottom to top of figure) which may require some consideration of the need to armour this part of the channel, although it may also be a facet of the digital terrain model, for which further analysis would be needed. The red areas above Dodd bottom are expected given the steep channel gradient

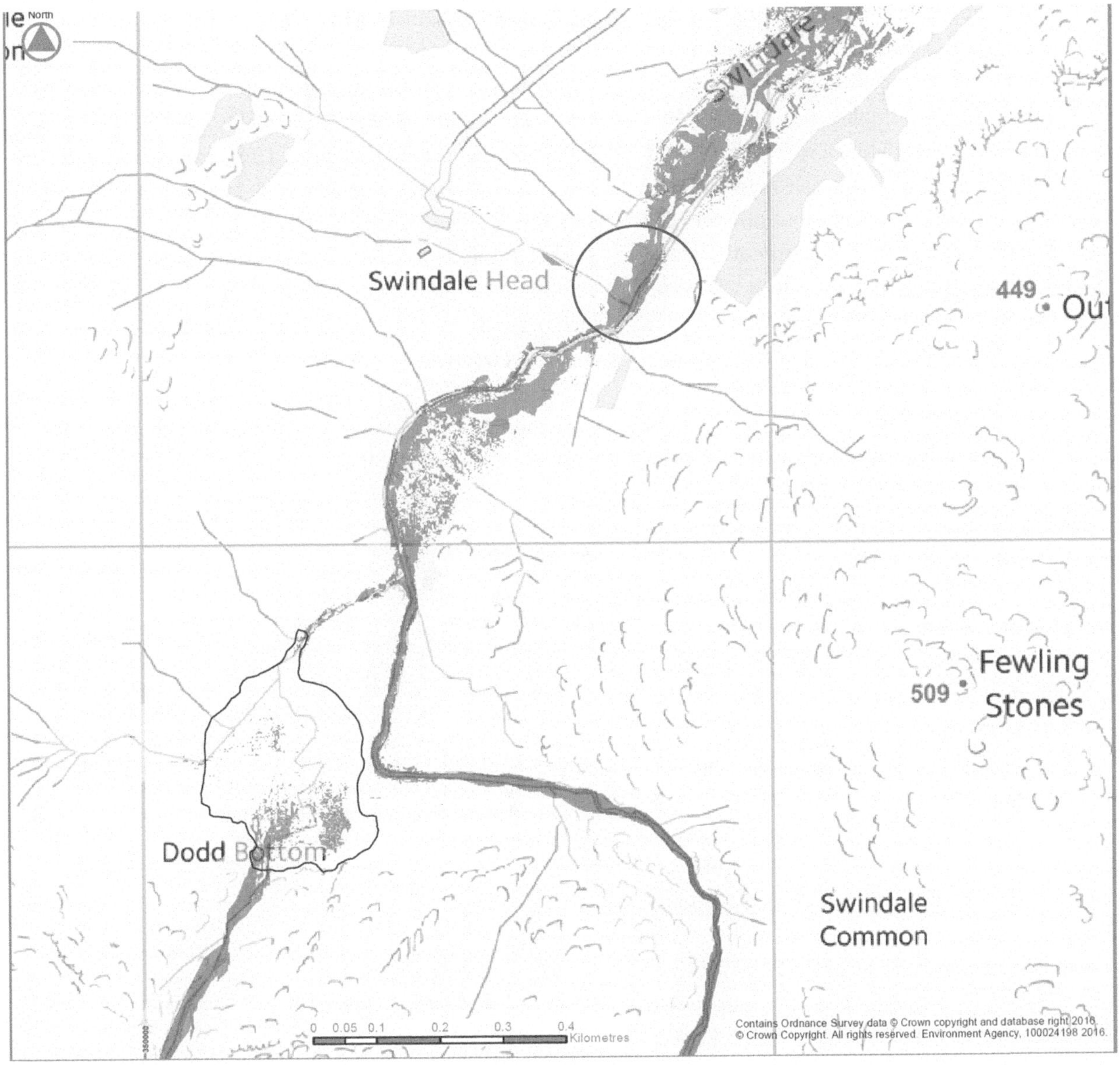

Figure 8 | Erodibility map based on shear stress analysis for the floodplain along Swindale beck, where erosion = red, deposition = yellow, and blue = transition. Please refer to the online version of this paper to see this figure in colour: http://dx.doi.org/10.2166/nh.2019.150.

and very rough bed. Identifying such issues is critical to the design of NFM schemes and their long-term performance.

CONCLUSIONS

The study demonstrates the feasibility of utilising the best aspects of two models, using Dynamic Topmodel for hillslope hydrology and JFlow or HEC-RAS2D for 2D channel–floodplain hydraulics, in order to simulate the detail of the processes that need to be represented with uncertainty in an NFM scheme evaluation. The credibility of the simulations of subsurface flows beneath hillslopes is increased with the use of Dynamic Topmodel for this component, while the credibility of the 2D floodplain hydraulics increased with the use of JFlow (or alternative 2D hydraulic models such as HEC-RAS2D). We have demonstrated that the approach is capable of a detailed simulation of interventions applied to floodplains (e.g., storage impoundments or channel–floodplain reconnection).

The approach has successfully advanced the coupled distributed hydrology and 1D hydraulics approach taken by Metcalfe *et al.* (2017) to 2D, permitting investigation of the upscaled impact of very small-scale ($2\,m^2$) NFM measures at the catchment scale ($15\,km^2$), whilst accounting for dynamic interactions in the drainage network. We have made use of the increasing versatility of different 2D hydrodynamic software packages to incorporate lateral inflows and hydraulic structures directly into a 2D mesh and allowing flexible mesh around key floodplain features such as embankments. Whilst there are numerous examples of broadscale coupling of hydrology and hydraulics models (Nguyen *et al.* 2015; Cea & Rodriguez 2016), these recent developments make integration of distributed hillslope-hydrological processes with distributed 2D hydrodynamic modelling considerably more feasible for a range of catchment investigations.

We have demonstrated that the approach can be used for identifying the quantum of change due to an isolated runoff attenuation feature, and also the uncertainty associated with this, given as: $1.8\,m^3\,s^{-1} \pm 1.4\,m^3\,s^{-1}$. This represents a $4\% \pm 2\%$ change at the flow gauge, which can help decision makers understand whether the intervention is worth investing in or re-engineering as a risk reduction strategy. In reality, further scenarios can be investigated, some diverting more of the flow from the main channel into Dodd Bottom, or through adding multiple features.

The approach will next be scaled up to the $210\,km^2$ Kent catchment, with 5 flow gauges and 72 km of watercourse as part of the NERC-funded Q-NFM project (https://www.lancaster.ac.uk/lec/sites/qnfm), to investigate the wider applicability of the approach. This larger catchment poses a computational problem, whereby it will no longer be feasible to simulate all of the behavioural Dynamic Topmodel scenarios using 2D hydraulic modelling at 2 m resolution, and a further sampling strategy will need to be devised. Since many of the hydraulic outputs are similar, we intend to record statistics of each of the inflow time series from Dynamic Topmodel for each reach (such as peak, volume, time of peak) and use these to distinguish between scenarios likely to result in different spatial behaviours. These can then be further constrained with local knowledge of particular flooding pathways during recent flooding.

ACKNOWLEDGEMENTS

This project has been supported by NERC grant NE/R004722/1 and has been a collaborative effort between JBA Consulting and Lancaster University in the UK. The Environment Agency is thanked for the use of rainfall and terrain data under licence CL77737MG, while Lee Schofield of the RSPB is thanked for the use of the additional terrain data and streamflow. Thanks also to members of the Q-NFM team at Lancaster, Trevor Page, Ann Kretzschmar and Rob Lamb for their helpful discussions. The ensemble Dynamic Topmodel simulations were undertaken by Peter Metcalfe, who sadly is no longer with us to see the results.

REFERENCES

Addy, S. & Wilkinson, M. 2016 An assessment of engineered log jam structures in response to a flood event in an upland gravel-bed river. *Earth Surface Processes and Landforms* **1**, 1658–1670. https://onlinelibrary.wiley.com/doi/full/10.1002/esp.3936

Aronica, G., Hankin, B. G. & Beven, K. J. 1998 Uncertainty and equifinality in calibrating distributed roughness coefficients in a flood propagation model with limited data. *Advances in Water Resources* **22** (4), 349–365.

Beven, K. J. 2006 A manifesto for the equifinality thesis. *Journal of Hydrology* **320** (1–2), 18–36.

Beven, K. & Binley, A. 1992 The future of distributed models: model calibration and uncertainty prediction. *Hydrological Processes* **6**, 279–298.

Beven, K. J. & Binley, A. 2014 GLUE: 20 years on. *Hydrological Processes* **28**, 5897–5918.

Beven, K. J. & Freer, J. 2001 A dynamic TOPMODEL. *Hydrological Processes* **15** (10), 1993–2011.

Blazkova, S. & Beven, K. J. 2009 A limits of acceptability approach to model evaluation and uncertainty estimation in flood frequency estimation by continuous simulation: Skalka catchment, Czech Republic. *Water Resources Research* **45**, W00B16.

Burgess-Gamble, L., Ngai, R., Wilkinson, M., Nisbet, T., Pontee, N., Harvey, R., Kipling, K., Addy, S., Rose, S., Maslen, S., Jay, H., Nicholson, A., Page, T., Jonczyk, J. & Quin, P. 2017 *Working with Natural Processes – Evidence Directory*. Environment Agency Project SC150005. https://assets.publishing.service.gov.uk/government/uploads/system/uploads/attachment_data/file/681411/Working_with_natural_processes_evidence_directory.pdf

Carrick, J., Shaiful Azman Bin Abdul Rahim, M., Adjei, C., Habib Hassan Ashraa Kalee, H., James Banks, S., Bolam, F. C.,

Campos Luna, I. M., Clark, B., Cowton, J., Freitas Nongando Domingos, I., Duba Golicha, D., Gupta, G., Grainger, M., Hasanaliyeva, G., Hodgson, D. J., Lopez-Capel, E., Magistrali, A. J., Merrell, I. G., Oikeh, I., Salim Othman, M., Kumari Ranathunga Ranathunga Mudiyanselage, T., Warren, C., Samuel, C., Sufar, E. K. H., Watson, P. A., Azwanida Binti Zakaria, N. N. & Stewart, G. 2018 Is planting trees the solution to reducing flood risks? *Journal of Flood Risk Management* **2018**, 1–10. DOI: 10.1111/jfr3.12484.

Cea, M. & Rodriguez, M. 2016 Two-dimensional coupled distributed hydrologic–hydraulic model simulation on watershed. *Pure and Applied Geophysics* **173**, 909. https://doi.org/10.1007/s00024-015-1196-5.

Clilverd, H. M., Thompson, J. R., Heppell, C. M., Sayer, C. D. & Axmacher, J. C. 2013 River-floodplain hydrology of an embanked lowland Chalk river and initial response to embankment removal. *Hydrological Sciences Journal* **58** (3), 1–24.

Dadson, S., Hall, J. W., Murgatroyd, A., Acreman, M., Bates, P., Beven, K., Heathwaite, L., Holden, J., Holman, I. P., Lane, S. N., O'Connell, E., Penning-Rowsell, E., Reynard, N., Sear, D., Thorne, C. & Wilby, R. 2017 A restatement of the natural science evidence concerning catchment-based natural flood management in the UK. *Proceedings of the Royal Society A* **473**, 20160706. http://dx.doi.org/10.1098/rspa.2016.0706.

Ghimire, S., Wilkinson, M. & Donaldson-Selby, G. 2015 Application of 1d and 2d Numerical Models For Assessing And Visualizing Effectiveness Of Natural Flood Management (NFM) Measures. In: *Proceedings of the International Conference on Hydroinformatics*.

Hankin, B., Metcalfe, P., Craigen, I., Page, T., Chappell, N., Lamb, R., Beven, K. & Johnson, D. 2017a Strategies for testing the impact of natural flood risk management measures. In: *Flood Risk Management* (T. V. Hromadka II & P. Rao, eds). InTechOpen, London. Available from: https://www.intechopen.com/books/flood-risk-management/strategies-for-testing-the-impact-of-natural-flood-risk-management-measures.

Hankin, B., Lamb, R., Craigen, I., Page, T., Chappell, N. & Metcalfe, P. 2017b *A whole catchment approach to improve flood resilience in the Eden. Winning entry to the Defra Flood Modelling Competition* 2016 [Internet]. 19-01-2017. Available from: https://consult.defra.gov.uk/water-and-flood-risk-management/flood-risk-management-modelling-competition/results/jba_defra_winning_entry_full_report.pdf

Holden, J., Kirkby, M. J., Lane, S. N., Milledge, D. G., Brookes, C. J., Holden, V. & McDonald, A. T. 2008 Overland flow velocity and roughness properties in peatlands. *Water Resources Research* **44**. doi:10.1029/2007wr006052.

Hollaway, M. J., Beven, K. J., Benskin, C. M. H., Collins, A. L., Evans, R., Falloon, P. D., Forber, K. J., Hiscock, K. M., Kahana, R., Macleod, C. J. A., Ockenden, M. C., Villamizar, M. L., Wearing, C., Withers, P. J. A., Zhou, J. G. & Haygarth, P. M. 2018 Evaluating a processed based water quality model on a UK headwater catchment: what can we learn from a 'limits of acceptability' uncertainty framework? *Journal of Hydrology* **558**, 607–624.

Hunter, N. M., Bates, P. D., Neelz, S., Pender, G., Villanueva, I., Wright, N. G., Liang, D., Falconer, R. A., Lin, B., Waller, S., Crossley, A. J. & Mason, D. C. 2008 Benchmarking 2d hydraulic models for urban flooding. *Proceedings of the Institute of Civil Engineers, Water Management* **161** (1), 13–30, Issue WM1.

Kjeldsen, T. R., Stewart, E. J., Packman, J. C., Folwell, S. & Bayliss, A. C. 2005 *Revitalisation of the FSR/FEH Rainfall-Runoff Method*. Defra R&D Technical Report FD1913/TR.

Lamb, R., Crossley, A. & Waller, S. 2009 A fast 2d floodplain inundation model. *Proceedings of the Institution of Civil Engineers: Water Management* **162** (6), 363–370.

Lane, S. N. 2017 Natural flood management. *Wiley Interdisciplinary Reviews: Water* **4** (3), e1211.

Lane, S. N. & Ferguson, R. I. 2005 Modelling reach-scale fluvial flows. In: *Computational Fluid Dynamics: Applications in Environmental Hydraulics* (P. D. Bates, S. N. Lane & R. I. Ferguson, eds). John Wiley & Sons Ltd., Chapter 10, pp. 217–269. Available from: https://onlinelibrary.wiley.com/doi/book/10.1002/0470015195

Liu, Y., Freer, J. E., Beven, K. J. & Matgen, P. 2009 Towards a limits of acceptability approach to the calibration of hydrological models: extending observation error. *Journal of Hydrology* **367**, 93–103. doi:10.1016/j.jhydrol.2009.01.016.

Metcalfe, P., Beven, K. & Freer, J. 2016 *dynatopmodel: Implementation of the Dynamic TOPMODEL Hydrological Model*. R package version 1.1. https://CRAN.R-project.org/package=dynatopmodel

Metcalfe, P., Beven, K., Hankin, B. & Lamb, R. 2017 A modelling framework for evaluation of the hydrological impacts of nature-based approaches to flood risk management, with application to in-channel interventions across a 29 km² scale catchment in the United Kingdom. *Hydrological Processes*. HYP-16-0365.R1. https://doi.org/10.1002/hyp.11140.

Metcalfe, P., Beven, K., Hankin, B. & Lamb, R. 2018 Simplified representation of runoff attenuation features within analysis of the hydrological performance of a natural flood management scheme. *Hydrology and Earth System Sciences* **22**, 2589–2605. https://doi.org/10.5194/hess-2017-398.

Nguyen, P., Thorstensen, A., Sorooshian, S., Hsu, K., AghaKouchak, A., Sanders, B., Koren, V., Cui, Z. & Smith, M. 2015 A high resolution coupled hydrologic–hydraulic model (HiResFlood-UCI) for flash flood modeling. *Journal of Hydrology* **541** (A), 401–420. http://dx.doi.org/10.1016/j.jhydrol.2015.10.047

Nicholson, A. R., Wilkinson, M. E., O'Donnell, G. M. & Quinn, P. F. 2012 Runoff attenuation features: a sustainable flood mitigation strategy in the Belford catchment, UK. *Area* **44** (4), 463–469. doi: 10.1111/j.1475-4762.2012.01099.x.

Office International de l'Eau 2013 Individual Natural Water Retention Measures: Floodplain Restoration and Management. 07.0330/2013/659147/SER/ENV.C1 for the Directorate General for Environment of the European

Commission. http://nwrm.eu/sites/default/files/nwrm_ressources/n3_-_floodplain_restoration_and_management.pdf

Pattison, I., Lane, S. N., Hardy, R. J. & Reaney, S. M. 2014 The role of tributary relative timing and sequencing in controlling large floods. *Water Resources Research* **50**, 5444–5458. doi:10.1002/2013WR014067.

Pitt, M. 2008 *The Pitt Review, 2008. Learning Lessons From the 2007 Floods*. June 2008. http://archive.cabinetoffice.gov.uk/pittreview/thepittreview/final_report.html

Reid, H. E., Williams, R. D., Brierley, G. J., Coleman, S. E., Lamb, R., Rennie, C. D. & Tancock, M. J. 2018 Geomorphological effectiveness of floods to rework gravel bars: insight from hyperscale topography and hydraulic modelling. *Earth Surface Processes and Landforms* https://doi.org/10.1002/esp.4521

RSPB 2017 Modelling Working with Natural Process in Swindale Beck. JBA Final Report. REF 2017s5854.

Shuttleworth, E. L., Evans, M. G., Pilkington, M., Spencer, T., Walker, J., Milledge, D. & Allott, T. E. H. 2019 Restoration of blanket peat moorland delays stormflow from hillslopes and reduces peak discharge. *Journal of Hydrology X* **2**, 100006. https://doi.org/10.1016/j.hydroa.2018.100006.

World Wildlife Fund 2016 *The Flood Green Guide. Natural and Nature-Based Flood Management: A Green Guide*. https://c402277.ssl.cf1.rackcdn.com/publications/1058/files/original/WWF_Flood_Green_Guide_FINAL.pdf?1495628174

First received 29 November 2018; accepted in revised form 24 June 2019. Available online 17 July 2019

Regional rainfall response to the North Atlantic Oscillation (NAO) across Great Britain

Harry West, Nevil Quinn IWA and Michael Horswell

ABSTRACT

The North Atlantic Oscillation (NAO) has been long studied as the primary teleconnection affecting the British and European climate. However, previous studies have focused on extremes or have been spatially and temporally limited. In recent years, our ability to predict the NAO has improved. Also, new research is emerging, suggesting that the NAO is a key driver of hydrological extremes. These factors mean that there is a renewed value in enhancing our understanding of how the NAO influences general rainfall patterns. In this study, we spatially analyse correlations between NAO indices and monthly rainfall data and the Standardised Precipitation Index. We also map mean monthly rainfall differences under NAO-positive and -negative conditions. Based on our results, we identify three main observations: (I) there is sensitivity in the rainfall patterns to the chosen NAO index; (II) there is a clear winter north/west and south/east divide in rainfall patterns; and (III) the NAO does have an effect on summer rainfall patterns, although the spatiality of these patterns is less distinctive than in winter. As far as we are aware, this is the first national scale, monthly NAO–rainfall analysis undertaken for a long period.

Key words | British climate, North Atlantic Oscillation, rainfall patterns, teleconnections, wet/dry continuum

Harry West (corresponding author)
Nevil Quinn IWA
Michael Horswell
Centre for Water, Communities and Resilience,
Department of Geography and Environmental Management,
University of the West of England,
Bristol,
UK
E-mail: *harry.west@uwe.ac.uk*

INTRODUCTION

Climate change is expected to significantly alter hydrometeorological and climatological processes and patterns in Great Britain (Garner *et al.* 2017; Kendon *et al.* 2018), and understanding the characteristics and impacts of wet and dry extreme events remains a challenge (Jones *et al.* 2014; Watts *et al.* 2015; Parry *et al.* 2016; Van Loon *et al.* 2016). Oceanic–atmospheric circulation interactions (also referred to as teleconnections) have a key influence on regional climate (Wilby *et al.* 1997; Rust *et al.* 2018). Given the close interconnectedness of the climate and hydrological system, floods and droughts are inherently driven by larger-scale meteorological processes and their interactions with local-scale catchment characteristics (Wilby & Quinn 2013; Van Loon & Laaha 2015; Barker *et al.* 2016; De Luca *et al.* 2017; Huang *et al.* 2017). Furthering our understanding of the influence of teleconnections on local and regional climate is therefore key in water and climate resilience planning.

Weather in Great Britain is highly variable, often fluctuating between wet and dry conditions (e.g. the very wet winter of 2013/2014 and the dry summer of 2018). The North Atlantic Oscillation (NAO) characterises some of the variability of the North Atlantic jet stream and has been acknowledged as the primary teleconnection affecting the British climate (Wilby *et al.* 1997; Hurrell *et al.* 2003). For example, Rodwell *et al.* (1999) suggested that the NAO is the single most important teleconnection influencing climate variability in the Northern Hemisphere, and Sweeney & O'Hare (1992) linked variations in precipitation to large-scale NAO fluctuations across Europe. More recent studies continue to emphasise the influence of the NAO on British and European climatic patterns (Comas-Bru & McDermott

doi: 10.2166/nh.2019.015

2014; Varouchakis *et al.* 2018). In a review of studies relating to hydro-meteorological signal control of the NAO, Rust *et al.* (2018) highlight the strong relationship (positive correlations) between the NAO and precipitation during winter months across Northern Europe.

The NAO can be defined in various ways but generally represents the sea level pressure (SLP) fluctuation between Iceland and the Azores that has been well defined in meteorology since the 19th century (Hurrell *et al.* 2003). Many studies use the NAO Index (NAOI) as a quantitative measure of the pressure gradient between Iceland and the Azores (e.g. Simpson & Jones 2014; Bonaccorso *et al.* 2015; Spencer & Essery 2016).

The NAO fluctuates between a positive and negative state; each state is known to produce characteristic climatic patterns over Great Britain and mainland Europe (Hall & Hanna 2018; Rust *et al.* 2018). A positive NAO (which represents stronger than the usual difference in the SLP between Iceland and the Azores) is generally associated with stormy and wet winter conditions as winds from the west dominate bringing warmer air and storms across the North Atlantic region. A negative NAO represents the reverse with a weaker than usual difference in the SLP. Winds from the east and north-east are more frequent, bringing with them cold air, while the adjusted position of the jet stream leads to weaker and less frequent storms. As a result, Europe is more likely to experience cold, calm and dry winters (Hurrell *et al.* 2003; Baker *et al.* 2017). Many studies have associated fluctuations of the NAO to precipitation patterns in Great Britain specifically (Figure 1).

The impact of the NAOI phase during winter and summer is known to differ significantly (Folland *et al.* 2009; del Rio *et al.* 2011; Sun & Wang 2012). While the strength of regional correlations with rainfall of the winter (December, January, February (DJF)) NAO Index ($NAOI_W$) and summer (June, July, August (JJA)) NAO Index ($NAOI_S$) vary across Great Britain, a general relationship is evident, particularly for the north/west of the country. There is a positive correlation between positive $NAOI_W$ values and winter precipitation, and a negative correlation between positive $NAOI_S$ values and summer precipitation. Conversely, negative $NAOI_S$ values are correlated with higher summer precipitation, and negative $NAOI_W$ values are associated with lower winter precipitation (Folland *et al.* 2009; Simpson & Jones 2014; Hall & Hanna 2018).

Earlier analyses of relationships between NAOI and precipitation are typically based on precipitation data from relatively few sites (e.g. Wilby *et al.* 1997; Fowler & Kilsby 2002) and short record lengths (e.g. Fowler & Kilsby 2002; Afzal *et al.* 2015). Studies have also tended to focus on extremes rather than more general wet/dry patterns (e.g.

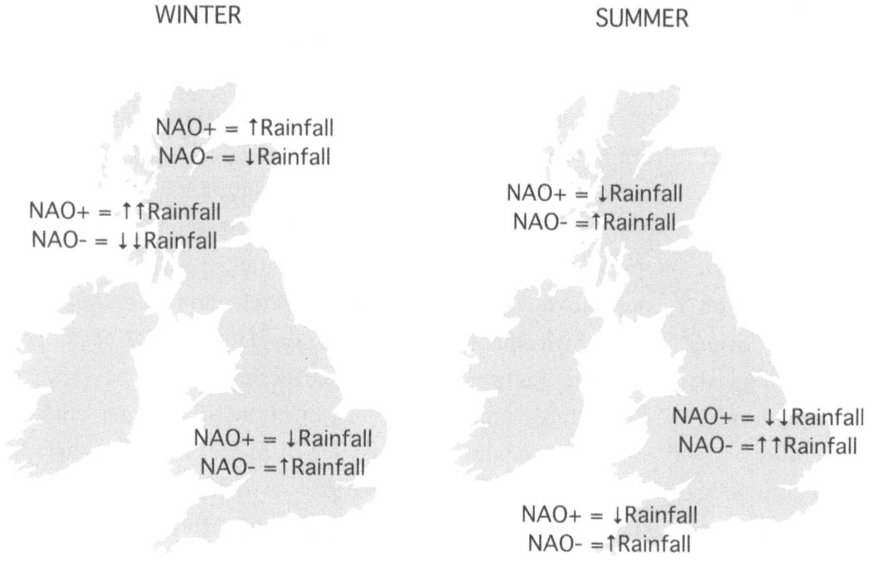

Figure 1 | Generalised NAO winter and summer rainfall patterns from previous studies (Wilby *et al.* 1997; Fowler & Kilsby 2002; Burt & Howden 2013; Kosanic *et al.* 2014; Simpson & Jones 2014; Afzal *et al.* 2015; Hall & Hanna 2018). A more detailed summary of these findings can be found in Table S1 in the online Supplemental Material.

Simpson & Jones 2014) and have had a clear emphasis on the relationship between the NAO and winter climate (e.g. Comas-Bru & McDermott 2014; Rust *et al.* 2018). This seems an understandable focus as intense and more frequent rainfall is typically associated with low pressure systems coming from the tropics – the movement of these systems being directly associated with the NAO phase (Wilby *et al.* 1997; Fowler & Kilsby 2002). While relationships between winter rainfall and the NAO are now well defined, the influence of the NAO on regional summer climates is less clear (Folland *et al.* 2009) and is limited to some evidence, suggesting negative correlations between the NAOI and summer precipitation in all regions apart from north-west Scotland (Hall & Hanna 2018).

The availability of nationally consistent gridded precipitation datasets, such as the Centre for Ecology and Hydrology (CEH) Gridded Estimates of Areal Rainfall (GEAR) dataset (Tanguy *et al.* 2016), now provides new opportunities for analysis of teleconnection drivers of rainfall. Similarly, the recent publication of the gridded Standardised Precipitation Index (SPI) time series (Tanguy *et al.* 2017) offers an additional advantage as the data are conveniently scaled in relation to relative wetness and dryness over a specified rainfall accumulation period (Hannaford *et al.* 2011). As a result, the SPI has been used to assess the spatial signature of teleconnections in rainfall in regions worldwide (Irannezhad *et al.* 2015; Kingston *et al.* 2015). The availability of these datasets is particularly important in helping develop a deeper understanding of the spatial structure of associations between North Atlantic teleconnections and precipitation, and any time-related/seasonal trends in this structure. Previous studies using similar spatially consistent data have tended to rely on UKCP09 estimates (e.g. Spencer & Essery 2016), which are known to have limitations/pre-conditions for estimates of summer rainfall (Met Office 2016). New understanding of the relationships between hydrological extremes and the NAO (Rust *et al.* 2019), coupled with improvements in long-range NAO prediction ability (Scaife *et al.* 2014; Smith *et al.* 2016; Baker *et al.* 2017; Weisheimer *et al.* 2017) and the potential for the NAOI to be used in hindcasting (Smith *et al.* 2019), also means that there is now a renewed value in developing a more complete understanding of how the NAO influences the spatial distribution of seasonal rainfall across Great Britain. Being able to infer more detailed regional rainfall responses in relation to predicted teleconnection behaviour would be of significant value in helping inform strategic responses to weather extremes (Palin *et al.* 2016; Bell *et al.* 2017; Clark *et al.* 2017).

In this study, we aim to examine the spatial distribution of rainfall (using both GEAR monthly rainfall estimates and the SPI) under positive and negative NAOI conditions. This represents the first monthly, nationally consistent spatial analysis undertaken for a long record (1899–2015, 117 water years) addressing the full wet/dry continuum.

METHODS

Data

To represent the NAO condition, the NAOI was used. This study makes use of two commonly used NAOI calculation methods: the station-based method ($NAOI_{ST}$) which calculates the index based on the fixed SLP station measured data, and the principal component analysis method ($NAOI_{PC}$) comprising a time series of the leading empirical orthogonal function of Atlantic SLP (Hurrell *et al.* 2003). Both datasets were obtained from the US National Centre for Atmospheric Research (https://climatedataguide.ucar.edu/climate-data) at a monthly interval for the water years October 1899–September 2015.

Monthly total rainfall values (1899–2015) were obtained from the CEH GEAR dataset (Tanguy *et al.* 2016). GEAR provides interpolated monthly estimates of total rainfall on a 1 km grid and has a relatively complete spatial coverage of Great Britain. The rainfall estimates are derived from the Met Office national database of observed precipitation. The GEAR dataset was selected due to its high spatial resolution and long record period (when compared to similar UK rainfall datasets). The SPI, as defined by Mckee *et al.* (1993), was also sourced from CEH at monthly intervals (1899–2015) with a 1-month rainfall accumulation period (Tanguy *et al.* 2017). This dataset is provided at 5 km gridded spatial resolution and is derived based on the monthly GEAR dataset. The SPI is calculated by fitting a gamma distribution (Stagge *et al.* 2015) to historical precipitation time series. The years 1961–2010 are used as the baseline for

SPI calculation. The SPI is a standardised index, with positive values representing wetter than normal conditions for a given period, and negative values representing drier than normal conditions. The SPI is normally distributed with a mean of 0 (i.e. near-normal conditions) and a standard deviation of 1. Theoretically, values can range from −5 (extremely dry) to +5 (extremely wet), although approximately 95% of values occur within the range of −2 to +2, and 68% within the range of −1 to +1 (Tanguy *et al.* 2017).

Regional rainfall/SPI and NAOI correlations

The Met Office Climate Districts are commonly used in similar studies (e.g. Wilby *et al.* 1997; Simpson & Jones 2014) and, as they represent areas of relatively homogeneous climate, were the chosen spatial unit of analysis for the correlation of regional mean monthly total rainfall/SPI and the NAOI. Mean monthly rainfall (based on GEAR data) and SPI were calculated by spatially averaging all gridded values within each climate district. The mean monthly rainfall/SPI for the study period are then correlated with the NAOI calculated using both the ST and PC method for Great Britain. Exploratory analysis established normality and linear relationships, so a Pearson correlation was deemed appropriate. In our analysis, we defined seasons as winter (DJF), spring (MAM), summer (JJA) and autumn (SON).

Calculation of the deviation of mean monthly rainfall and mean SPI-1

The monthly NAOI was classified into a state of positive, negative or neutral NAO phase. In light of the known limitations of the $NAOI_{ST}$ method of calculation, especially in the summer months (see Discussion), only the $NAOI_{PC}$ method was used for further analysis. The NAO phase was defined as half the standard deviation plus/minus the long-term mean (after Berton *et al.* 2017). The NAO-positive phase was calculated to be NAOI >0.502 and -negative phase <−0.503. Months with a $NAOI_{PC}$ between these values represent an NAO neutral state. Each month in the GEAR and SPI-1 dataset was categorised using these values into being in either an NAO-positive, NAO-negative or NAO-neutral state (Figure 3). For each month, all datasets for a given phase were averaged on a cell-by-cell basis to produce a mean dataset for each month under each NAO condition. For the GEAR data, the NAO-positive/-negative mean monthly datasets were then subtracted from the NAO neutral dataset for that month. This produced a final dataset to show deviation in mean monthly rainfall in the given phase compared to when NAO is neutral.

The deviations in mean monthly rainfall and mean SPI-1 values were then spatially averaged using the Met Office Climate Districts. In order to examine how similar each region's response to the NAO was over the study period, the mean deviation from neutral NAO conditions/mean monthly SPI of each district was then correlated against values from the other districts. Regions which are positively correlated suggest areas where rainfall displays a similar NAO response and vice versa for negatively correlated regions.

RESULTS

Regional mean correlations

Figure 2 shows the results of the regional mean SPI correlation analysis with $NAOI_{ST}$ and $NAOI_{PC}$. Similar seasonal variations were produced using both rainfall measures (SPI and monthly GEAR data). Positive correlations suggest high SPI/GEAR values (i.e. high rainfall) when under NAO-positive conditions and vice versa for NAO-negative. Negative correlations are indicative of low SPI/GEAR values (i.e. low rainfall) when NAO is positive and vice versa for NAO negative. Significant positive correlations between $NAOI_{ST}$ and monthly rainfall are found in the west of Great Britain, and particularly in the north, for all months between October and April. While significant correlations persist in 'Scotland North' and 'Scotland West' in spring (MAM), they are largely absent in England and Wales. June is characterised by a marked gradient with most of England showing significant negative correlations. July and August are characterised by a notable lack of significant correlations between $NAOI_{ST}$ and the rainfall measures, with July largely having weak positive correlations and August a combination of weak positive and negative correlations. September to November sees a change in the correlation strength; such as that by

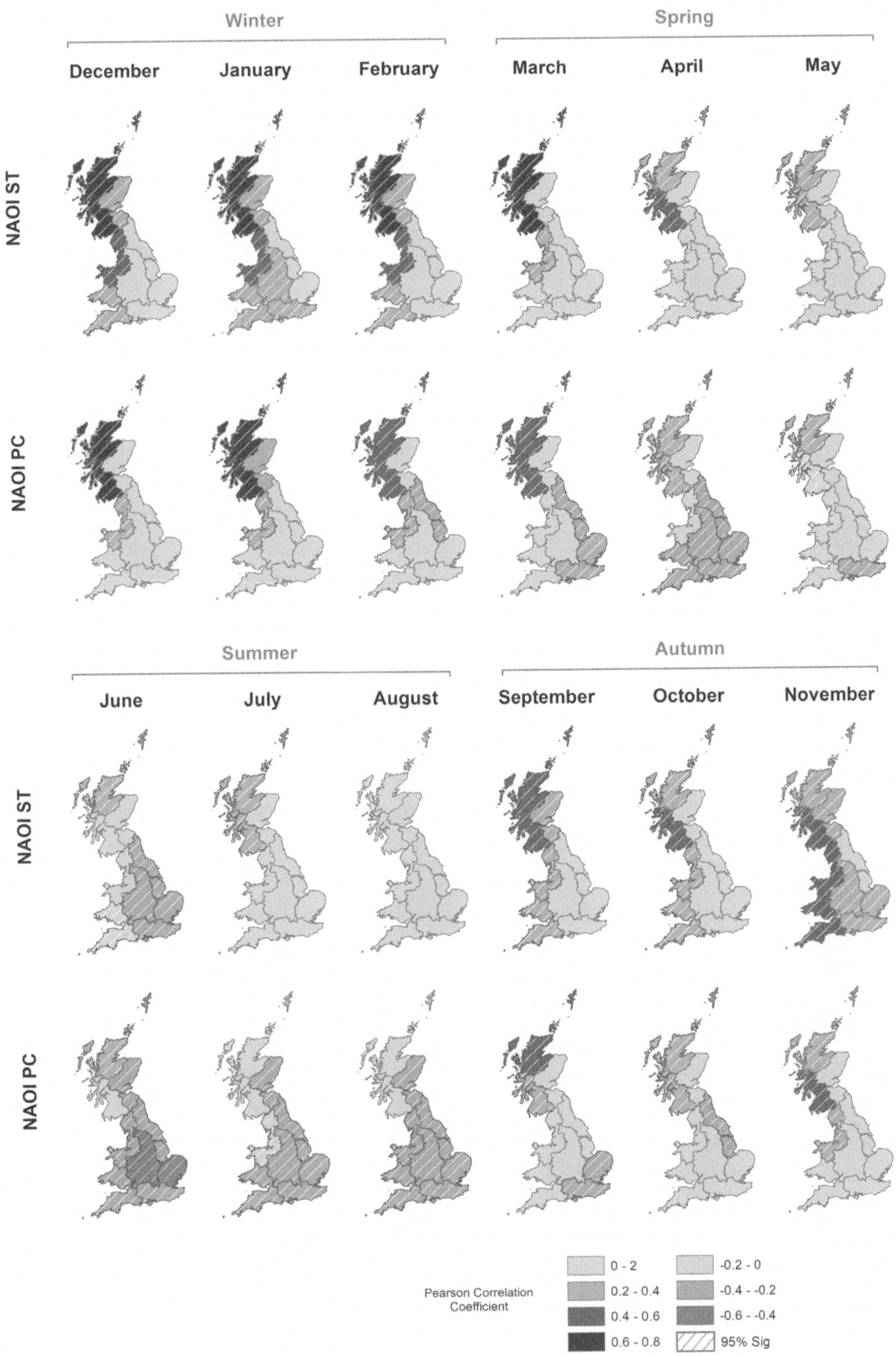

Figure 2 | Correlations between $NAOI_{ST}$ and $NAOI_{PC}$ with regional mean SPI-1. The two NAOI methods show similar patterns in the winter months (DJF), with strong positive correlations in the North. However, note that the significant differences observed between the two NAOIs in the summer months where stronger negative correlations are observed using $NAOI_{PC}$. Tabular data for this analysis can be found in the online Supplemental Material (Tables S2 and S3).

November, 'England East and North East' is the only region not showing a significant positive correlation.

$NAOI_{PC}$ produces similar spatial patterns to those described above; significant positive correlations are found for the north and west for the months of September to May. During this time, the central and southern regions fluctuate between positive and negative correlations (of varying significance). $NAOI_{PC}$ correlations vary from those gained through using $NAOI_{ST}$ in spring (MAM), where more pronounced negative correlations are present in the central and southern regions. This signal is enhanced moving into the summer months where all of England, Wales and 'Scotland East' see significant negative correlations between SPI and $NAOI_{PC}$.

NAO-phase definition

Figure 3 shows the results of the NAO-phase classification process using the method of Berton *et al.* (2017) on the $NAOI_{PC}$ time series. Based on our definition of phases, the NAO was in a positive or negative phase for approximately 53% of the time between October 1899 and September 2015. The NAO-phase occurrence and intensity were clearly more pronounced during the winter (DJF) rather than summer (JJA) months, especially for NAO-positive.

Spatial patterns of deviation in mean monthly rainfall and mean SPI-1 under the NAO phase

Figure 4 shows the NAO phase-dependent deviation in mean monthly rainfall (calculated as the difference from mean monthly rainfall under NAO neutral conditions). Clear regional and seasonal differences are observed in both NAO-positive and -negative phases. The most notable deviations in rainfall occur in the north of Great Britain during the winter months (DJF) (in particular in Scotland North and Scotland West). These regions receive significantly more or less rainfall under NAO-positive or -negative conditions, respectively. These conditions are inverted in the south/east of Great Britain, with these regions seeing decreases/increases under NAO-positive/-negative, however to a lesser extent compared to the north. This creates a strong winter spatial signal in rainfall under the two NAO conditions, with the north/west and south/east showing clear regional differences. The magnitude of the deviation is also different in terms of enhanced wetness/dryness; NAO-positive tends to produce wetter conditions than NAO-negative does dry. This suggests that the enhancing effect on rainfall is more marked than the dampening effect. Only small changes in rainfall under NAO conditions were found in southern Great Britain throughout winter. Moving into spring (MAM), this spatial pattern is largely retained, although the magnitude of the deviation in rainfall decreases with time.

Significant differences in rainfall are also observed during the summer months (JJA). However, the relative increases/decreases in rainfall under NAO-positive/-negative are broadly inverted when compared to the winter months (DJF). The NAO spatial signature now seems more apparent in the southern and central regions, with NAO-positive resulting in notably drier conditions, and NAO-negative producing wetter conditions (up to approximately ± 30 mm), most notably in July and August in South West England and South Wales. The clear north/west and south/east spatial divide in rainfall patterns observed in the winter months becomes less pronounced, with the central and southern regions of Great Britain being more homogeneous in their rainfall response, and much of the country displaying a similar rainfall deviation to these regions. In the autumn months (SON), the patterns in rainfall deviation start to invert again becoming more similar to those described above in winter and the regional differences in the NAO rainfall response become more discernible.

Figure 5 shows the comparable analysis undertaken using mean monthly SPI-1 data. As with the deviation in monthly rainfall analysis discussed above, notable seasonal and spatial patterns in wet/dry conditions are present when NAO is in either a positive or negative phase. During the winter (DJF), the north has significantly high/low SPI values, representing wetter than normal/drier than normal conditions under positive/negative NAO. Mean SPI-1 values follow a similar winter spatial pattern to the monthly deviation analysis – a clear north/west and south/east difference in the NAO response. While during the summer months (JJA) this pattern is broadly inverted and more homogeneous spatially so that the central and southern regions experience the most notable drier/wetter conditions under NAO positive/negative, with the rest of the country experiencing similar conditions.

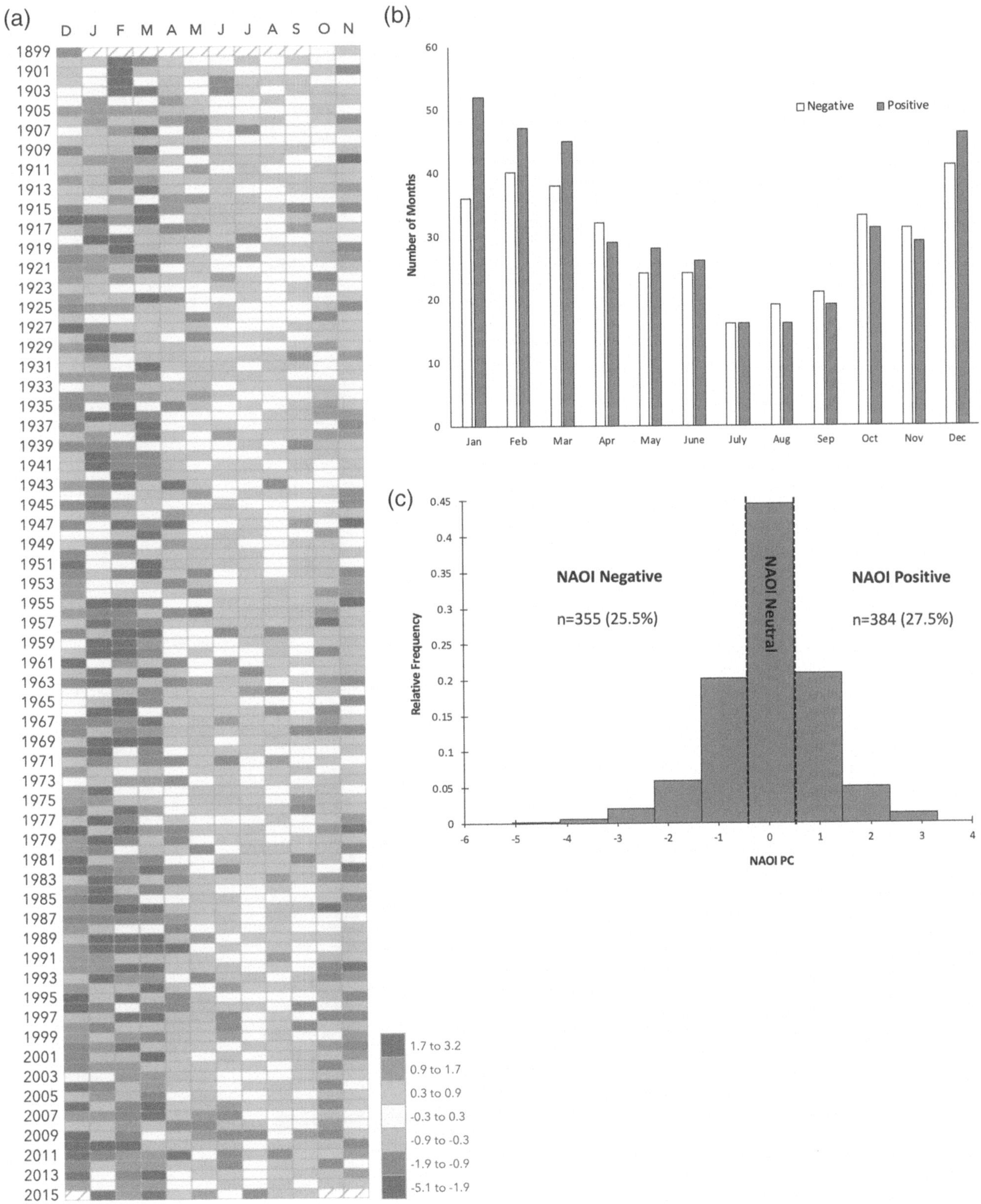

Figure 3 | (a) Heatmap showing the temporal distribution of NAO-phase occurrence and intensity, (b) the frequency of occurrence of each NAO phase per month and (c) the distribution of $NAOI_{PC}$ classification over the study period (1899–2015).

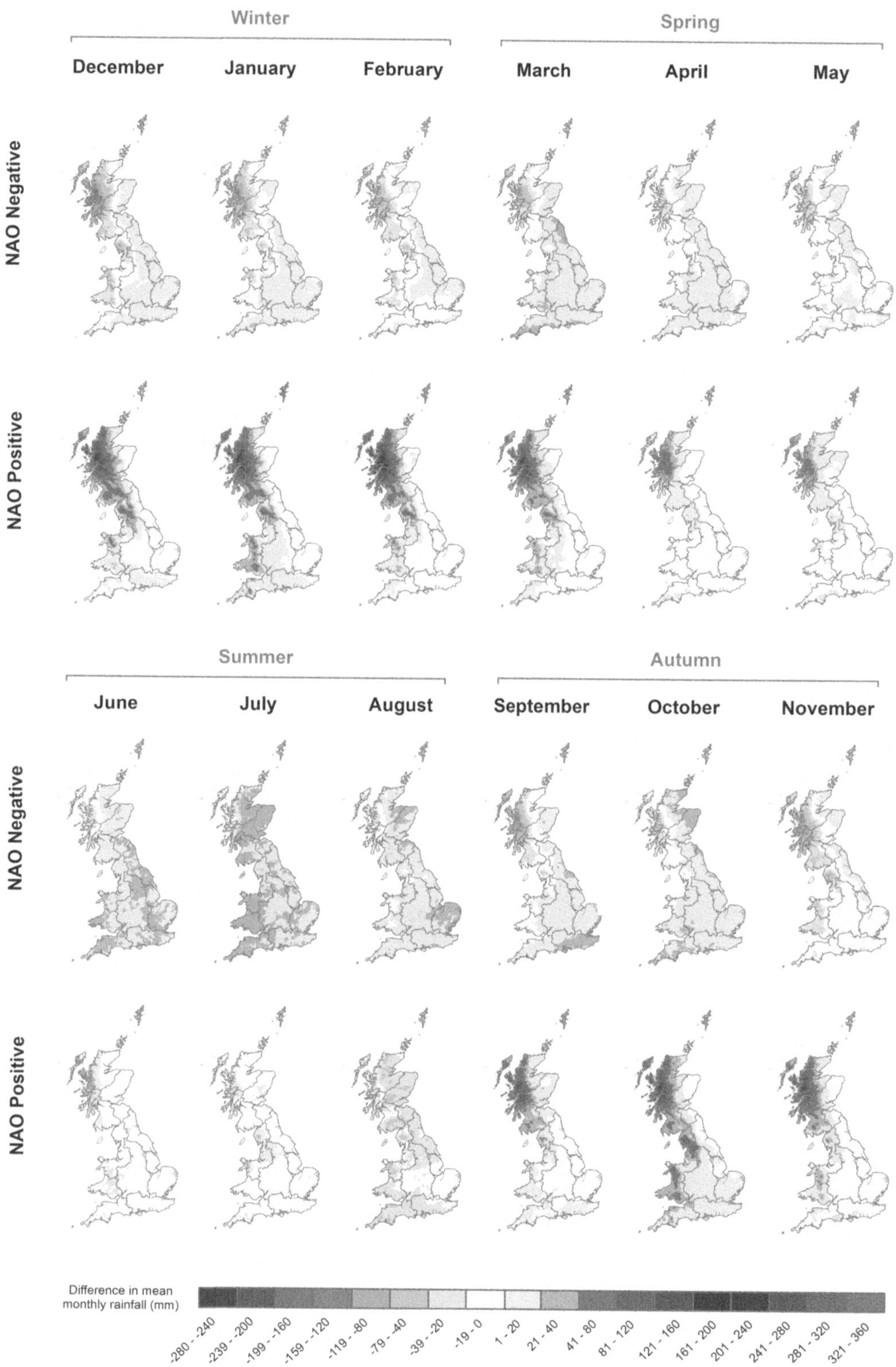

Figure 4 | Difference in mean monthly rainfall values (mm) under the NAO phase for the period October 1899–September 2015. Note the significant increase/decrease in rainfall in the north/west during the winter months and the wetter/drier conditions observed during the summer months under NAO-positive/negative (GEAR data from Tanguy *et al.* 2016). Tabular data for this analysis can be found in the online Supplemental Material (Table S4).

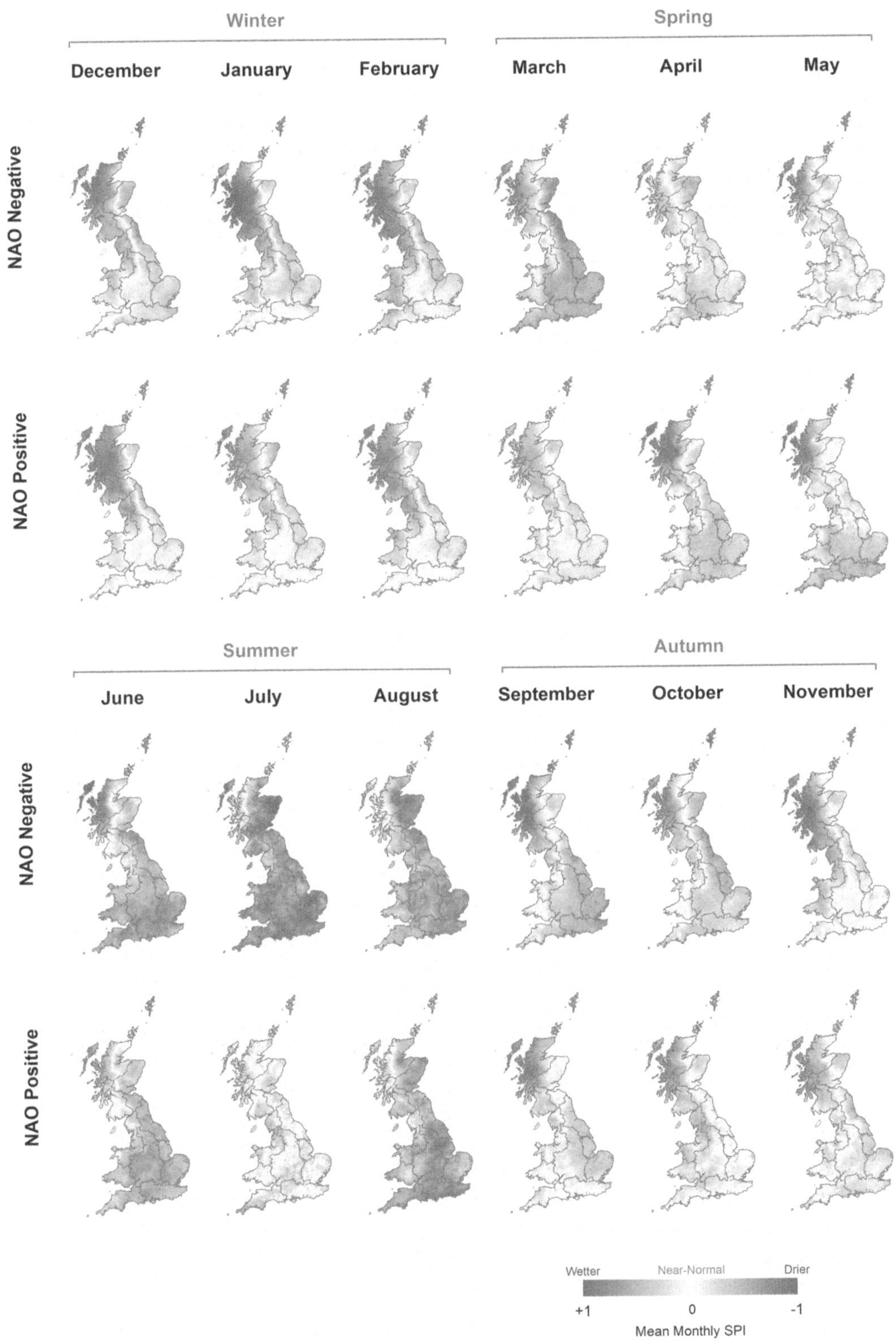

Figure 5 | Mean monthly SPI-1 under NAO conditions for the period October 1899–September 2015. Note the significantly wetter/drier conditions in the north/west during the winter months and the wetter/ drier conditions observed during the summer months under NAO positive/negative. (SPI data from Tanguy *et al.* 2017). Tabular data for this analysis can be found in the online Supplemental Material (Table S5).

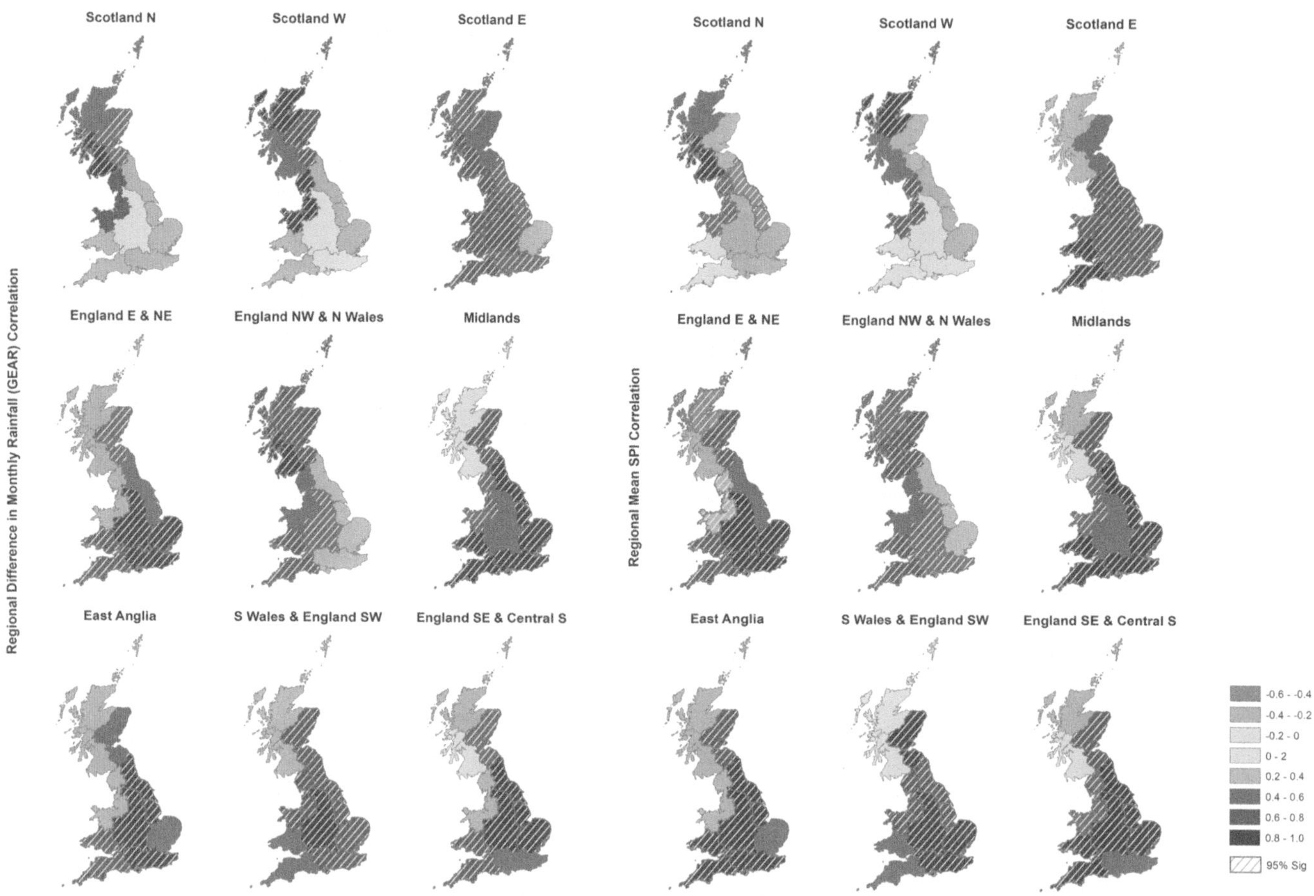

Figure 6 | Spatial representation of Pearson correlation coefficient values for assessing the similarity of rainfall response across the nine Met Office Climate Districts. Note the positive correlations of the north-western/south-eastern regions and the negative correlations between Scotland and the Southern/Central regions. Block grey areas indicate the region used as the independent variable in the correlation analysis. Tabular data for this analysis can be found in the online Supplemental Material (Tables S6 and S7).

Figure 6 shows the results of the correlation analysis to assess the similarity of each region's response to NAO-positive/-negative conditions (derived using both the mean monthly rainfall deviation data and mean SPI-1 data). Regions which show positive correlations in their spatial mean rainfall deviation or SPI-1 respond to the NAO in a similar way (i.e. they are wetter or drier by a similar magnitude), while regions which are negatively correlated show where differences occur in the regional response to NAO fluctuation. Similar spatial patterns in correlation are present using both mean monthly rainfall deviation (from the GEAR dataset) and mean monthly SPI-1.

'Scotland North' and 'Scotland West' are strongly positively correlated at the 95% confidence level; these regions are also positively correlated with 'England North West and North Wales'. This suggests that these regions respond to the NAO in a similar manner. A clear grouping is also evident in the positive correlations around the South East of Great Britain, with 'East Anglia' and 'England South East and Central South' showing similar responses. Based on these regional groupings of positive correlations, the north/west and south/east divide present in the previous analysis is clearly shown in the correlation maps, most notably in the southern regions' negative correlation with 'Scotland North' and 'Scotland West'.

However, the clear spatial signature is only evident in the correlations between certain regions. Some regions, such as 'Scotland East', 'South Wales and England South West', and the 'Midlands', show consistent significant positive correlations with almost all other regions. This suggests that these regions respond in a similar way to the NAO and tend to follow the general pattern in deviation/mean SPI-1 as the rest of the country.

DISCUSSION

This research sets out to examine the spatiality of rainfall patterns under NAO-positive and -negative conditions using nationally consistent, high-resolution datasets (Tanguy *et al.* 2016, 2017). This extends previous NAO rainfall studies based in Great Britain as it considers the full wet/dry continuum, rather than just weather extremes, across a full year, rather than just the winter months, and is not restricted in terms of available data as we use spatially consistent gridded datasets rather than analysis based on selected stations/regions. A range of methods were implemented to achieve this, and based on the convergence of evidence in the results, we identify three main observations: (I) the sensitivity of spatial rainfall analysis to the chosen NAOI calculation method; (II) the clear north/west and south/east divide in rainfall signatures under NAO-positive and -negative conditions during the winter months; and (III) the NAO does appear to have some influence over summer (JJA) rainfall, although the spatiality of these patterns is less distinctive when compared to winter. These will now be discussed in turn.

The first observation relates to the use of different NAOI measures; in our study, namely the decision to use $NAOI_{ST}$ or $NAOI_{PC}$. Pearson correlation analysis was undertaken using both SPI-1 and monthly GEAR data against both NAOI measures (Figure 2). While the two measures generally produced similar spatial patterns in correlation strength during the winter months, notable differences were observed during spring (MAM) and summer (JJA). $NAOI_{PC}$ produced significant strong negative correlations in the southern and central regions during this period, while the $NAOI_{ST}$ method produced much weaker correlations which were not significant. This implies that if $NAOI_{ST}$ is used, then under positive NAO much of the country is wetter than average, and drier under NAO negative. However, if $NAOI_{PC}$ is used, then this pattern is the opposite. Therefore, had the later analysis to examine rainfall spatiality been undertaken using the $NAOI_{ST}$ method, it is likely that the results would have been far less conclusive as those gained from the $NAOI_{PC}$ method (Kosanic *et al.* 2014).

The poor representation of summer NAO has been noted as a key limitation of the station-based method of NAOI calculation (Pokorná & Huth 2015). The main limitation is the mobility of the dipoles of the NAO across any given year (Jung *et al.* 2003; Beranová & Huth 2008). During the summer months, the main pressure 'action points' of the NAO (the Azores high and Icelandic low) move away from the position of the *in situ* monitoring stations, measurements from which are used to calculate $NAOI_{ST}$, due to a combination of factors including the East Atlantic and Scandinavian teleconnections (Hurrell & Van Loon 1997; Moore *et al.* 2013) which have also been attributed to UK meteorological patterns (Comas-Bru & McDermott 2014; Zubiate *et al.* 2017). Based on our correlation analysis and the potential misrepresentation of summer rainfall patterns under NAO conditions, we support the conclusions of Hurrell & Deser (2009) who suggest that there is no unique or universally accepted way to define the NAO, and those of Pokorná & Huth (2015) who advise caution in the selection of NAO representative indices and recommend non-station-based methods of calculation for summer NAO representation due to circulation spatial variability.

The spatial analysis of rainfall patterns under NAO-positive and -negative phase informs our second and third main observations. The accepted general understanding of NAO impact on rainfall across Great Britain during the winter months is to associate winter storms with NAO-positive and colder/drier winters with NAO-negative indices (Visbeck *et al.* 2001). Our analysis of both deviations in mean monthly rainfall and mean SPI-1 values shows a clear north/west and south/east spatial divide in response to NAO during the winter months, implying a less spatially generalisable response than suggested above. The north/west of Great Britain, in particular, sees rainfall patterns significantly alter under different NAO phases. Under NAO positive, these regions see notable increases in monthly rainfall, while under NAO negative there is a notable decrease. Although to a smaller magnitude, the opposite response is observed in the southern and eastern regions. This clear winter spatial pattern in rainfall under NAO positive/negative conditions is also evident in the regional correlation analysis, where the northern and western regions are positively correlated with each other, but negatively correlated with the southern and eastern regions. This winter spatial pattern has been observed in other studies over shorter timescales using station-based measurements of rainfall, thus indicating some long-term consistency in this pattern (Wilby *et al.* 1997; Fowler & Kilsby 2002; Burt & Howden 2013; Simpson & Jones 2014). Although this winter pattern coincides with upland/lowland topographic patterns, which may enhance rainfall under NAO-positive conditions (Burt & Howden 2013), the notable dry conditions under NAO-negative suggest that the interaction between NAO and local characteristics is far more complex. Having an understanding of the generalisable winter NAO rainfall pattern is significant given that in recent years notable advances have been made in our ability to predict winter NAO-phase months in advance (Baker *et al.* 2017; Hall *et al.* 2017; Weisheimer *et al.* 2017). For example, being able to say with some degree of confidence that the NAO will remain in a negative condition over the winter months will allow for early warning water shortage systems in Scotland to be triggered to ensure consistent supply in the upcoming dry period.

The clear spatial pattern during winter which reverses under the two different NAO conditions is likely associated with the location of storm tracks due to altered jet stream location (Visbeck *et al.* 2001). Long-term storm track analysis (mid-Holocene) suggests that a NAO-positive phase is characterised by a stronger and more northerly/easterly storm track across the Atlantic (Trigo 2006; Brayshaw *et al.* 2010). This NAO-driven pattern is evident in both precipitation and wind speed records (Burningham & French 2013).

Our analysis shows that the NAO winter spatial pattern continues into spring (MAM), gradually decreasing in discernibility over time; this is explainable by the SLP anomalies associated with winter NAO tending to persist into spring across the Atlantic region (Herceg-Bulić & Kucharski 2014). This change in the NAO rainfall signature over time brings into focus our final main observation. Moving into summer (JJA), the spatial pattern of rainfall under both positive and negative NAO is far less distinctive and more homogenous (Figures 4 and 5). Some regional differences are observable, with the central and southern regions showing the most notable deviation from normal conditions. However, the general direction of this deviation under NAO-positive and -negative conditions is broadly inverted during summer when compared to winter. NAO-positive now produces relatively homogenous dry conditions, and NAO-negative produces wetter conditions. Alongside the NAO, it is likely that high SPI values (high rainfall) are associated with convective storms during the summer months (Kendon *et al.* 2014). As these systems are small in their spatial influence, this may partly explain the dispersed summer rainfall patterns in Figures 4 and 5.

CONCLUSION

As the primary teleconnection affecting the British climate, the NAO has been well researched (Wilby *et al.* 1997; Fowler & Kilsby 2002; Burt & Howden 2013). However, many studies have focused on winter climate and weather extremes and have been based on a limited number of *in situ* measurements/regions. This study sets out to examine the spatial signatures of NAO phase-dependent precipitation, using nationally consistent datasets. Through correlation and spatial analysis, we draw attention to three observations regarding regional response in rainfall patterns to the NAO: (I) the sensitivity of any analysis to the chosen NAOI; (II) the clear north/west and south/east divide in rainfall signatures during the winter months; and (III) the NAO does appear to have some influence over summer rainfall, although the spatiality of these patterns is less distinctive when compared to winter patterns.

Our results extend the understanding of the NAO's influence on rainfall patterns in Great Britain and suggest that, now more than ever, there is potential to improve our predictive ability of teleconnections and precipitation, allowing for better informed water management decisions. This is important given the potential future impacts of climate change on hydro-meteorological conditions in Great Britain (Watts *et al.* 2015). In relation to seasonal forecasting and climate projections, this work shows some potential and adds to the growing evidence finding value of including the NAOI in model simulations (Baker *et al.* 2017; Smith *et al.* 2019).

It is important to note that our work uses gridded rainfall from the CEH-GEAR dataset, which over the period shows variability in the number of rainfall stations used for interpolation. Therefore, an avenue for future research would be to repeat similar analysis using more temporally consistent Met Office MIDAS (Met Office Integrated Data Archive System) rainfall. The use of the new CEH SPI data (Tanguy *et al.* 2017) also has potential for future research in this area, due to the calculation of multiple accumulation periods ranging from 3 to 24 months. Future research could utilise these longer periods to establish whether there is any lagged effect of the NAO on rainfall patterns. It is also important to note that the NAO cannot explain all of the rainfall spatio-temporal patterns in the UK. NAO rainfall spatial signatures have been found to show variable levels of consistency over long records (West *et al.* 2019), and extreme events have been found to occur due to the interaction of multiple teleconnections (Rust *et al.* 2019). There is therefore still a future need to consider the NAO's interaction with and the general influence of other teleconnections on weather and climate in Great Britain to further improve our predictive capability of seasonal rainfall patterns (Comas-Bru & McDermott 2014; Hall & Hanna 2018).

ACKNOWLEDGEMENTS

We thank the two anonymous reviewers, whose constructive comments and suggestions have allowed us to strengthen this manuscript and clarify our argument.

SUPPLEMENTARY MATERIAL

The Supplementary Material for this paper is available online at https://doi.org/10.2166/nh.2019.015.

REFERENCES

Afzal, M., Gagnon, A. S. & Mansell, M. G. 2015 Changes in the variability and periodicity of precipitation in Scotland. *Theoretical and Applied Climatology* **119**, 135–159.

Baker, L. H., Shaffrey, L. C. & Scaife, A. A. 2017 Improved seasonal prediction of UK regional precipitation using atmospheric circulation. *International Journal of Climatology* **38** (51), 437–453.

Barker, L. J., Hannaford, J., Chiverton, A. & Svensson, C. 2016 From meteorological to hydrological drought using standardised indicators. *Hydrology and Earth System Science* **20**, 2483–2505.

Bell, V. A., Davies, H. N., Kay, A. L., Brookshaw, A. & Scaife, A. A. 2017 A national-scale seasonal hydrological forecast system: development and evaluation over Britain. *Hydrology and Earth System Sciences* **21**, 4681–4691.

Beranová, R. & Huth, R. 2008 Time variations of the effects of circulation variability modes on European temperature and precipitation in winter. *International Journal of Climatology* **28** (2), 139–158.

Berton, R., Driscoll, C. T. & Adamowski, J. F. 2017 The near-term prediction of drought and flooding conditions in the northeastern United States based on extreme phases of AMO and NAO. *Journal of Hydrology* **553**, 130–141.

Bonaccorso, B., Cancelliere, A. & Rossi, G. 2015 Probabilistic forecasting of drought class transitions in Sicily (Italy) using Standardised Precipitation Index and North Atlantic Oscillation Index. *Journal of Hydrology* **526**, 136–150.

Brayshaw, D. J., Hoskins, B. & Black, E. 2010 Some physical drivers of changes in the winter storm tracks over the North Atlantic and Mediterranean during the Holocene. *Philosophical Transactions of the Royal Society of London A: Mathematical, Physical and Engineering Sciences* **366** (1931), 5185–5223.

Burningham, H. & French, J. 2013 Is the NAO winter index a reliable proxy for wind climate and storminess in northwest Europe? *International Journal of Climatology* **33**, 2036–2049.

Burt, T. & Howden, N. 2013 North Atlantic Oscillation amplifies orographic precipitation and river flow in upland Britain. *Water Resources Research* **49**, 3504–3515.

Clark, R. T., Bett, P. E., Thornton, H. E. & Scaife, A. A. 2017 Skilful seasonal predications for the European energy industry. *Environmental Research Letters* **12**, 024002.

Comas-Bru, L. & McDermott, F. 2014 Impacts of the EA and SCA patterns on the European twentieth century NAO – winter climate relationship. *Quarterly Journal of the Royal Meteorological Society* **140**, 354–363.

De Luca, P., Hillier, J. K., Wilby, R. L., Quinn, N. W. & Harrigan, S. 2017 Extreme multi-basin flooding linked with extra-tropical cyclones. *Environmental Research Letters* **12**, 114009.

del Rio, S., Herrero, L., Pinto-Gomes, C. & Penas, A. 2011 Spatial analysis of mean temperature trends in Spain over the period 1961–2006. *Global Planetary Change* **78** (1–2), 65–75.

Folland, C. K., Knight, J., Linderholm, H. W., Fereday, D., Ineson, S. & Hurrell, J. W. 2009 The summer North Atlantic Oscillation: past, present and future. *Journal of Climate* **22**, 1082–1103.

Fowler, H. J. & Kilsby, C. G. 2002 Precipitation and the North Atlantic Oscillation: a study of climatic variability in Northern England. *International Journal of Climatology* **22**, 843–866.

Garner, G., Hannah, D. & Watts, G. 2017 Climate change and water in the UK: recent scientific evidence for past and future change. *Progress in Physical Geography* **41** (2), 154–170.

Hall, R. J. & Hanna, E. 2018 North Atlantic circulation indices: links with summer and winter temperature and precipitation and implications for seasonal forecasting. *International Journal of Climatology* **38** (S1), 660–677.

Hall, R. J., Jones, J. M., Hanna, E., Scaife, A. A. & Erdélyi, R. 2017 Drivers and potential predictability of summer North Atlantic polar front jet variability. *Climate Dynamics* **48**, 3869–3887.

Hannaford, J., Lloyd-Hughes, B., Keef, C., Parry, S. & Prudhomme, C. 2011 Examining the large-scale spatial coherence of European drought using regional indicators of precipitation and streamflow deficit. *Hydrological Processes* **25**, 1146–1162.

Herceg-Bulić, I. & Kucharski, F. 2014 North Atlantic SSTs as a link between wintertime NAO and the following spring climate. *Journal of Climate* **27** (1), 186–201.

Huang, S., Li, P., Huang, Q., Leng, G., Hou, B. & Ma, L. 2017 The propagation from meteorological to hydrological drought and its potential influence factors. *Journal of Hydrology* **547**, 184–195.

Hurrell, J. W. & Deser, C. 2009 North Atlantic climate variability: the role of the North Atlantic Oscillation. *Journal of Marine Systems* **78**, 28–41.

Hurrell, J. W. & Van Loon, H. 1997 Decadal variations in climate associated with the North Atlantic Oscillation. *Climatic Change* **36** (3–4), 301–326.

Hurrell, J. W., Kushnir, Y., Ottersen, G. & Visbeck, M. 2003 An overview of the North Atlantic Oscillation. In: *The North Atlantic Oscillation: Climate Significance and Environmental Impact, AGU Geophysical Monograph Series, 134* (J. W. Hurrell, Y. Kushnir, G. Ottersen & M. Visbeck, eds). American Geophysical Union, Washington, 1–35. doi:10.1029/134GM01

Irannezhad, M., Torabi Haghighi, A., Chen, D. & Kløve, B. 2015 Variability in dryness and wetness in central Finland and the role of teleconnection patterns. *Theoretical and Applied Climatology* **122** (3–4), 471–486.

Jones, M. R., Blenkinsop, S., Fowler, H. J. & Kilsby, C. G. 2014 Objective classification of extreme rainfall for the UK and updated estimates of trends in regional extreme rainfall. *International Journal of Climatology* **34**, 751–765.

Jung, T., Hilmer, M., Ruprecht, E., Kleppek, S., Gulev, S. K. & Zolina, O. 2003 Characteristics of the recent eastward shift of interannual NAO variability. *Journal of Climate* **16**, 3371–3382.

Kendon, E. J., Roberts, N. M., Fowler, H. J., Roberts, M. J., Chan, S. C. & Senior, C. A. 2014 Heavier summer downpours with climate change revealed by weather forecast resolution model. *Nature Climate Change* **4**, 570–576.

Kendon, M., McCarthy, M., Jevrejeva, S., Matthews, A. & Legg, T. 2018 State of the UK climate 2017. *International Journal of Climatology* **38** (S2), 1–35.

Kingston, D. G., Stagge, J. H., Tallaksen, L. M. & Hannah, D. M. 2015 European-scale drought: understanding connections between atmospheric circulation and meteorological drought indices. *Journal of Climate* **28** (2), 505–516.

Kosanic, A., Harrison, S., Anderson, K. & Kavcic, I. 2014 Present and historical climate variability in South West England. *Climatic Change* **124**, 221–237.

McKee, T. B., Doesken, N. J. & Kleist, J. 1993 The relationship of drought frequency and duration to time scales. In: *Eighth Conference on Applied Climatology*, January 17–22, 1993, Anaheim, California.

Met Office 2016 *Is UKCP09 Still and Appropriate Tool for Adaption Planning? Land Projections*. Available from: http://ukclimateprojections.metoffice.gov.uk/media.jsp?mediaid=88738&filetype=pdf (accessed 03 May 2018).

Moore, G. W. K., Renfrew, I. A. & Pickart, R. S. 2013 Multidecadal variability of the North Atlantic Oscillation. *Journal of Climate* **26** (8), 2453–2466.

Palin, E. J., Scaife, A. A., Wallace, E., Pope, E. C. D., Arribas, A. & Brookshaw, A. 2016 Skillful seasonal forecasts of winter disruption to the UK transport system. *Journal of Applied Meteorology and Climatology* **55**, 325–344.

Parry, S., Prudhomme, C., Wilby, R. L. & Wood, P. J. 2016 Drought termination: concept and characterisation. *Progress in Physical Geography* **40** (6), 743–767.

Pokorná, L. & Huth, R. 2015 Climate impacts of the NAO are sensitive to how the NAO is defined. *Theoretical and Applied Climatology* **119** (3–4), 639–652.

Rodwell, M. J., Rowell, D. P. & Folland, C. K. 1999 Oceanic forcing of the wintertime North Atlantic Oscillation and European climate. *Nature* **398**, 320–323.

Rust, W., Holman, I., Corstanje, R., Bloomfield, J. & Cuthbert, M. 2018 A conceptual model for climatic teleconnection signal control on groundwater variability in Europe. *Earth-Science Reviews* **177**, 164–174.

Rust, W., Holman, I., Bloomfield, J., Cuthbert, M. & Corstanje, R. 2019 Understanding the potential of climate teleconnections to project future groundwater drought. *Hydrology and Earth System Sciences Discussions*. doi:10.5194/hess-2019-119.

Scaife, A. A., Arribas, A., Blockley, E., Brookshaw, A., Clark, R. T., Dunstone, N., Eade, R., Fereday, D., Folland, C. K., Gordon, M., Hermanson, L., Knight, J. R., Lea, D. J., MacLachlan, C., Maidens, A., Martin, M., Peterson, A. K., Smith, D., Vellinga, M., Wallace, E., Waters, J. & Williams, A. 2014 Skillful long-range prediction of European and North American winters. *Geophysical Research Letters* **41**, 2514–2519.

Simpson, I. R. & Jones, P. D. 2014 Analysis of UK precipitation extremes derived from Met Office Gridded Data. *International Journal of Climatology* **34**, 2438–2449.

Smith, K., Stringer, N., Tanguy, M., Eastman, M., Dadson, S. & Knight, J. 2019 Improving winter ensemble streamflow predicting in the UK using North Atlantic Oscillation analogues. In: *European Geosciences Union General Assembly*, April 7–12, 2019, Vienna, Austria.

Smith, D. M., Scaife, A. A., Eade, R. & Knight, J. R. 2016 Seasonal to decadal prediction of the winter North Atlantic Oscillation: emerging capability and future prospects. *Quarterly Journal of the Royal Meteorological Society* **142**, 611–617.

Spencer, M. & Essery, R. 2016 Scottish snow cover dependence on the North Atlantic Oscillation index. *Hydrology Research* **47** (3), 619–629.

Stagge, J. H., Tallaksen, L. M., Gudmundsson, L., Van Loon, A. F. & Stahl, K. 2015 Candidate distributions for climatological drought indices (SPI and SPEI). *International Journal of Climatology* **35** (13), 4027–4040.

Sun, J. & Wang, H. 2012 Changes of the connection between the summer North Atlantic Oscillation and the East Asian summer rainfall. *Journal of Geophysical Research* **117**, D08110.

Sweeney, J. C. & O'Hare, G. P. 1992 Geographical variations in precipitation yields and circulation types in Britain and Ireland. *Transactions of the Institute of British Geographers* **17** (4), 448–463.

Tanguy, M., Dixon, H., Prosdocimi, I., Morris, D. G. & Keller, V. D. J. 2016 Gridded estimates of daily and monthly areal rainfall for the United Kingdom (1890–2015) [CEH GEAR]. *NERC Environmental Information Centre*. doi:10.5285/33604ea0-c238-4488-813d-0ad9ab7c51ca.

Tanguy, M., Fry, M., Svensson, C. & Hannaford, J. 2017 Historic gridded standardized precipitation index for the United Kingdom 1862–2015 (generated using gamma distribution with standard period 1961–2010) v4. *NERC Environmental Information Centre*. doi:10.5285/233090b2-1d14-4eb9-9f9c-3923ea2350ff.

Trigo, I. F. 2006 Climatology and interannual variability of storm-tracks in the Euro-Atlantic sector: a comparison between ERA-40 and NCEP/NCAR reanalyses. *Climate Dynamics* **26** (2–3), 127–143.

Van Loon, A. F. & Laaha, G. 2015 Hydrological drought severity explained by climate and catchment characteristics. *Journal of Hydrology* **526**, 3–14.

Van Loon, A. F., Stahl, K., Di Baldassarre, G., Clark, J., Rangecroft, S., Wanders, N., Gleeson, T., Van Dijk, A. I. J. M., Tallaksen, L. M., Hannaford, J., Uijlenhoet, R., Teuling, A. J., Hannah, D. M., Sheffield, J., Svoboda, M., Verbeiren, B., Wagener, T. & Van Lanen, H. A. J. 2016 Drought in a human-modified world: reframing drought definitions, understanding, and analysis approaches. *Hydrology and Earth System Sciences* **20**, 3631–3650.

Varouchakis, E. A., Corzo, G. A., Karatzas, G. P. & Kotsopoulou, A. 2018 Spatio-temporal analysis of annual rainfall in Crete, Greece. *Acta Geophysica* **66** (3), 319–328.

Visbeck, M. H., Hurrell, J. H., Polvani, L. & Cullen, H. M. 2001 The North Atlantic Oscillation: past, present and future. *Proceedings of the National Academy of Sciences of the United States of America* **98** (23), 12876–12877.

Watts, G., Hannah, D. M. & Watkinson, A. 2015 Introduction to the special issue on the impact of climate change on water in the UK. *Progress in Physical Geography* **39** (1), 3–5.

Weisheimer, A., Schaller, N., O'Reilly, C., MacLeod, D. A. & Palmer, T. 2017 Atmospheric seasonal forecasts of the twentieth century: multi-decadal variability in predictive skills of the winter North Atlantic Oscillation (NAO) and their potential value for extreme event attribution. *Quarterly Journal of the Royal Meteorological Society* **143**, 917–926.

West, H., Quinn, N. & Horswell, M. 2019 A space-time geostatistical approach to exploring the stationarity of North Atlantic Oscillation driven wet/dry conditions in Great Britain. In *European Geosciences Union General Assembly*, April 7–12, 2019, Vienna, Austria.

Wilby, R. L. & Quinn, N. W. 2013 Reconstructing multi-decadal variations in fluvial flood risk using atmospheric circulation patterns. *Journal of Hydrology* **487**, 109–121.

Wilby, R. L., O'Hare, G. & Barnsley, N. 1997 The North Atlantic Oscillation and British Isles climate variability. *Weather* **52** (9), 266–276.

Zubiate, L., McDermott, F., Sweeney, C. & O'Malley, M. 2017 Spatial variability in winter NAO – wind speed relationships in western Europe linked to concomitant states of the East Atlantic and Scandinavian patterns. *Quarterly Journal of the Royal Meteorological Society* **143**, 552–562.

First received 8 February 2019; accepted in revised form 9 May 2019. Available online 25 May 2019

Insights into rainfall undercatch for differing raingauge rim heights

Katie Muchan and Harry Dixon

ABSTRACT

The measurement of rainfall has a long history, but despite its apparent simplicity it is difficult to quantify accurately. The common installation of raingauges with rims above the ground surface results in a difference between the rainfall caught and the amount reaching ground level, termed undercatch. The UK standard installation of raingauges is for their rim to be sited at 0.305 m above the ground; however, the use of weighing gauges installed at a minimum rim height of 1 m has increased in recent years. The installation of these weighing raingauges raises complex questions of homogeneity in rainfall data across space and time. Here, we investigate the impact of these changes using field trials of commonly deployed UK raingauges at a site in south-east England. This paper discusses the results of the trial, exploring the variation in and potential drivers of undercatch with differing gauge sitings. With varying standards for gauge heights around the world and new rainfall measurement technologies coming to the market all the time, improved understanding of undercatch is needed to inform evolving operational practices and explore the possibility of developing catch correction algorithms to remove arising inhomogeneity in precipitation datasets.
Key words | hydrometry, precipitation, rainfall, raingauge, undercatch

Katie Muchan (corresponding author)
Harry Dixon
UK Centre for Ecology & Hydrology,
Wallingford,
Oxfordshire OX10 8BB,
UK
E-mail: *katmuc@ceh.ac.uk*

INTRODUCTION

Accurate measurement of rainfall amount is crucial for many areas of hydrology, including water balance studies, flow forecasting, modelling and water resource assessments (Tian *et al.* 2007; Looper *et al.* 2012; Stisen *et al.* 2012). Rainfall has been measured since as early as the fourth century and there are currently many different types of gauge in use around the world, although manual storage gauges read by observers on a daily or monthly basis form a large part of the UK's long-established rainfall observational network (Strangeways 2010). This network is augmented with automatic recording gauges which are used to measure rainfall at a finer temporal resolution, essential for uses such as flood risk modelling and hazard warning systems (Tapiador *et al.* 2012). However, despite a long history of observations in the UK and around the world, it is widely acknowledged that there are many errors (both random and systematic) encountered when measuring rainfall whether manual or automatic gauges are used.

Irrespective of gauge type, a common issue in the measurement of rainfall is wind-induced undercatch where, due to the deformation of wind and increased turbulence above the gauge rim, raindrops are deflected away from the collecting orifice meaning less rain is recorded in a gauge mounted above the ground than would reach ground level (Rodda & Dixon 2012). This undercatch effect has been investigated extensively by field intercomparisons (e.g. Sevruk *et al.* 2009; Chubb *et al.* 2015; Pollock *et al.* 2018) and computational fluid dynamics modelling

doi: 10.2166/nh.2019.024

(e.g. Nespor & Sevruk 1999; Colli *et al.* 2018). In addition, accuracy is found to be affected by rainfall intensity, most notably in tipping-bucket gauges where counting errors have been observed (Molini *et al.* 2005). For the users of precipitation data, understanding the impact of this issue on the homogeneity of records across space and time is made more complex due to variations between different types of gauges and the way in which they are sited.

While there are a set number of gauge types available (Strangeways 2010), many of which have been subject to extensive intercomparison studies (e.g. Lanza & Vuerich 2009), there is a significant variation in the way that gauges are installed around the world. The World Meteorological Organization (WMO) recommends that the height of the gauge rim should be as low as possible but high enough to prevent splashing in from the ground surface (WMO 2008). A number of options exist to minimise the impact of increasing wind velocities with gauge height, including installing a shield to reduce the velocity around the gauge (Benning & Yang 2005), installing a turf wall (Essery & Wilcock 1991) or installing a gauge of a more aerodynamic shape (Strangeways 2004; Sieck *et al.* 2007; Colli *et al.* 2018). However, the most effective approach is to install a gauge within a pit to ensure the rim is at ground level preventing the body of the gauge from creating an obstacle to the wind (Rodda 1968). Extensive trials took place in the 1960s in Wallingford (Oxfordshire, UK) to assess the impact of the shape and size of pit grids, and now a European standard exists for the design of reference raingauge pits (CEN 2010) which can be used to evaluate wind effects or to conduct comparison against other reference raingauges. The installation of gauges in pits to create a reference rainfall series means that the rainfall recorded in a gauge above the ground surface may be corrected to the value reaching the surface (Allerup & Madsen 1980; Essery & Wilcock 1991; Mekonnen *et al.* 2015; Kochendorfer *et al.* 2017).

Tipping-bucket raingauges (TBRs) are one of the most popular automatic gauges in use globally but have many issues (Habib *et al.* 2001; Ciach 2003; Michaelides *et al.* 2009). TBRs are subject to random errors (due to clogging/blockages or mechanical failures (Sevruk 1996; Upton & Rahimi 2003)) and systematic errors (e.g. counting errors where rainfall is underestimated during high-intensity events as water is lost in the tipping movement of the buckets (Molini *et al.* 2005)). TBRs (and manual storage gauges) are also particularly unreliable in measuring solid precipitation, as it has to melt before it is recorded (Savina *et al.* 2012). Other technology exists in the form of weighing raingauges (WRGs), and although they are not deployed widely (Tapiador *et al.* 2012), their use is increasing globally (e.g. Sevruk & Chvíla 2005; Gray & Toucher 2019). Compared with TBRs, they can more easily provide a greater sensitivity of precipitation measurement (e.g. to 0.01 mm) as no moving buckets are required, they can minimise evaporation and wetting losses due to the lack of a funnel and have demonstrated improved snow measurement capabilities in some field trials (when heated due to reduced evaporation losses from a smaller heated area (e.g. Savina *et al.* 2012)).

For many years, the UK raingauge network comprised a mixture of storage and tipping-bucket gauges. However, due to the advantages mentioned above and the reduced maintenance and calibration requirements shown by a trial conducted at two sites in Scotland and Wales (Grust & Stewart 2012), WRGs have, in many cases, been deployed as replacements for TBRs in the UK since 2013. The Ott Pluvio gauge has been widely chosen as the WRG to replace TBRs and have been installed with their rims 1 m above the ground surface compared with the 0.305 m commonly used for TBRs and manual storage gauges in the UK.

Observation networks have always evolved as measurement practices and technologies are developed and changes in gauge type and/or siting such as that seen in the UK in recent years. However, these changes can potentially create unknown inhomogeneity in time series without extensive periods of dual running and/or specific intercomparison studies (Sevruk 1996; Savina *et al.* 2012). While comparisons between the different models of raingauge currently deployed in the UK have been conducted by others (Colli *et al.* 2018), and this study does not seek to repeat such work, the difference in shape and rim height between gauges raises the potential for increased amounts of wind-induced undercatch by the newly deployed WRGs. In this study, we explore undercatch across the different UK dominant raingauge types and common sitings and attempt to isolate the impact of wind-induced errors by comparing observations from each gauge with ground-level installations of the same make and model.

STUDY DESIGN

Site location

The meteorological station at the Centre for Ecology & Hydrology in Wallingford, Oxfordshire (UK) has been in operation since 1962. The site is located on the floodplain of the River Thames in south-east England at a height of 47 m above sea level. The wind direction is dominated by a west/south-west direction, and the mean annual precipitation (calculated using 1969–2018 manual daily pit gauge data) is 611 mm. The site contains equipment for 09:00 GMT daily manual observations of temperature, wind, precipitation and sunshine and forms part of the UK Climatological Observation Network for the UK Met Office (https://www.metoffice.gov.uk/public/weather/climate-network/#?tab=climateNetwork) (ID 5558). These measurements are replicated by an Automatic Weather Station (AWS) in the same compound, recording data at 15-min intervals.

Instrumentation

For the purpose of this study, the station contains seven rain-gauges (Figure 1) – two UK Met Office MK2 daily storage gauges, two Casella tipping-bucket gauges and three Ott Pluvio weighing gauges (Table 1). One gauge of each type is installed at the UK standard 0.305 m rim height and one Ott Pluvio at 1 m, reflecting the common siting used across the UK for such gauges in recent years. Each gauge is spaced to minimise the aerodynamic effects of each other, and no gauge is sited in a way which convenes the 26° exposure angle stated in the British Standard (BS7843-2:2012). Both TBRs have 0.2 mm buckets and have been calibrated throughout the operating period using volumetric calibration to an intensity of 10 mm/h. The Ott Pluvio gauges were calibrated and checked for drift in weighing capability (using known weights) throughout and no adjustments were required.

One gauge of each type is installed with a rim height at ground level (shown with an * in Table 1 and Figure 1) by siting them in separate 1.4 m × 1.4 m pits, 0.305 m deep (for storage and tipping-bucket gauges) and 1 m deep (for weighing gauges), with a 1.2 m × 1.2 m anti-splash metal grid on the surface, complying with the European Standards (CEN 2010) and WMO recommendations (WMO 2008). All analysis for gauges installed with their rim above ground level was done by comparison with the relevant reference pit gauges of the same types. This set-up removes the influence of changing gauge technologies from the trial and isolates the effect of undercatch. Wind speed is recorded by a cup anemometer mounted at 2 m.

Data preparation

The data and results presented here reflect the period where all three gauge types were in operation (August 2015–August 2018). All sub-daily rainfall data were quality controlled using the other gauge heights and types. Records identified as being anomalous (e.g. due to the Pluvio incorrectly recording rainfall during high temperatures when the collecting bucket is empty) were removed. Daily and monthly totals were then checked for any large discrepancies, but none were found. Any periods of missing data (e.g. due to instrument calibration or telemetry failure) were checked against other gauge types. For the study period, there were four large (>24-h) missing data periods, two of which were within periods of no rainfall and two coincided with small (<1 mm) daily rainfall totals. Therefore, with minimal impact of missing data on the analysis, all data were included. Daily and monthly totals were calculated over the UK standard 09:00–09:00 hydrological day. Seasonal figures were calculated for UK winter (December–February), spring (March–May), summer (June–August) and autumn (September–November) using monthly totals.

In order to investigate the potential driving factors behind any undercatch, an event dataset was created using the 15-min resolution reference Pluvio gauge (0.000 m) data against which the 0.305 and 1 m Pluvio data could be compared. Although the Pluvio provides measurements at 1-min resolution this was not used in this study, as operationally within the UK 15-min data are used for reporting and analysis. In addition, wind speed measurements from the Wallingford station were only available at 15-min intervals for direct comparison with the rainfall. The definition of rainfall event can vary depending on the purpose of a given study, but events are generally based on a period of accumulated rainfall above a set threshold

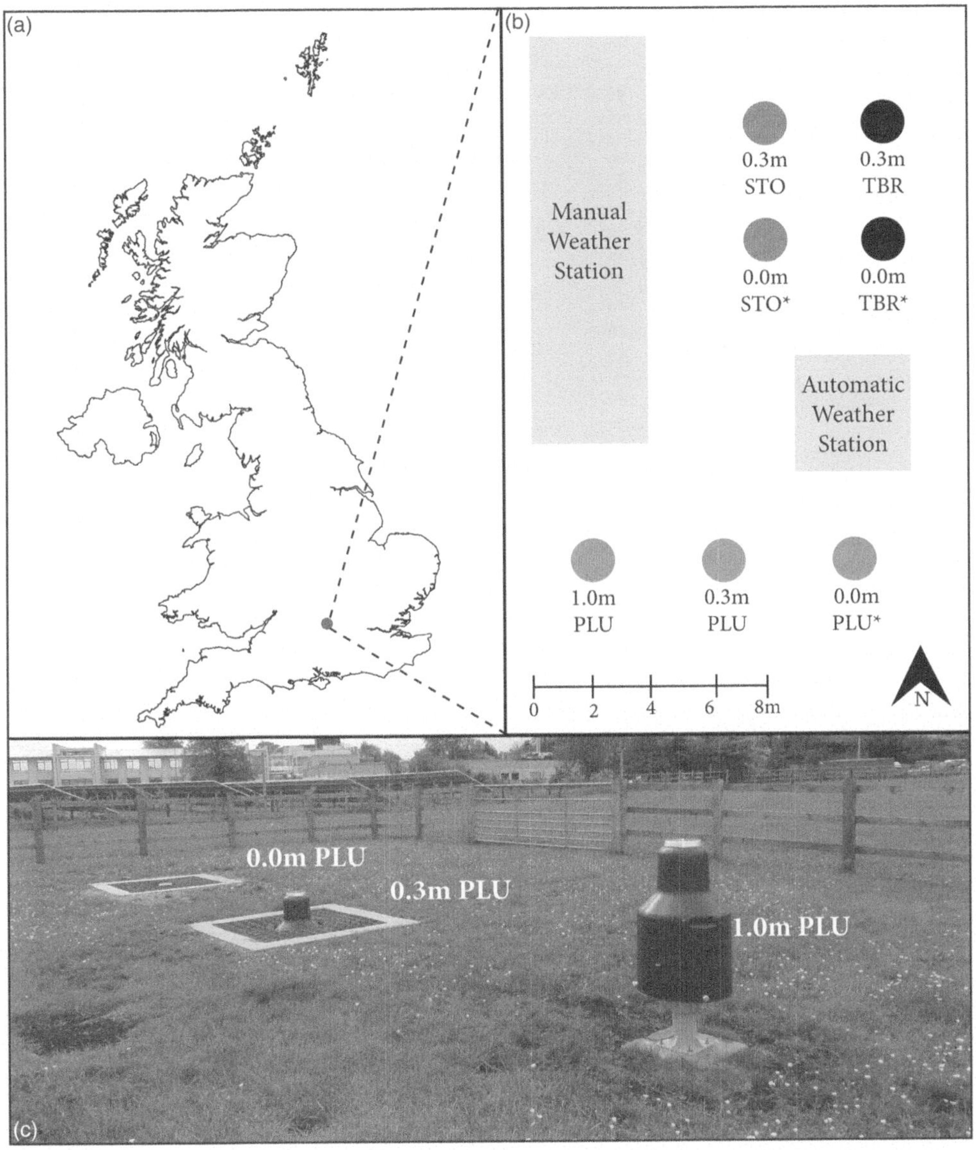

Figure 1 | (a) Location of meteorological station at the Centre for Ecology & Hydrology in Wallingord (UK); (b) site plan with reference gauges used in the study shown with an * (gauge size is not to scale shown) and (c) photograph of Ott Pluvio installation.

separated by a given period of time. This period is known as the 'minimum inter-event time' (MIT) and in a review by Dunkerley (2008) covering a wide range of geographical locations lengths of between 0.25 and 24 h were found to be in use, with the majority between 6 and 8 h. In southern England, rainfall events are a mixture of frontal events (long MIT) and convective storms (small MIT) (https://www.metoffice.gov.uk/climate/uk/regional-climates/so#rainfall). For this study, event datasets were produced and analysed for 1-, 2-, 3- and 4-h MITs. A 3-h MIT was chosen, as it was found to provide a good balance to include both convective and frontal events. In addition, an accumulation threshold of 1 mm (recorded in each of the three weighing gauges) was used, which generated a dataset of 288 events. Snowfall was present in six of these events, and as snow only represents a small proportion of the dataset and the climatology of the site is such that snowfall is rarely observed, these events were excluded creating a dataset of 282 events for analysis. For each event, the characteristics listed in Table 2 and the percentage catch for each Pluvio

Table 1 | Metadata (gauge type, sitings height, the temporal resolution of data collected and date of installation) of the raingauges used in this study

Abbreviation	Gauge type	Gauge rim height (m)	Temporal resolution	Year of installation
0.0 m STO*	Storage	0.000	Daily	1969
0.3 m STO	Storage	0.305	Daily	1962
0.0 m TBR*	Tipping bucket	0.000	15 min	2002
0.3 m TBR	Tipping bucket	0.305	15 min	2011
0.0 m PLU*	Weighing	0.000	15 min	2015
0.3 m PLU	Weighing	0.305	15 min	2015
1.0 m PLU	Weighing	1.000	15 min	2015

*The reference gauges for each type.

Table 2 | Metadata for characteristics derived from the Pluvio event dataset – name, unit, minimum, mean and maximum recorded in the 0.0 m PLU

Event characteristic	Unit	Minimum	Mean	Maximum
Duration	h	0.50	6.40	36.25
Rainfall total	mm	1.06	5.82	41.97
Average wind speed (when raining)	m/s	0.12	2.22	6.30
Average event intensity (rainfall total divided by event duration)	mm/h	0.33	1.54	10.99

installed above the ground compared with the reference gauge were calculated.

RESULTS

Monthly undercatch analysis

To assess the degree to which undercatch is observed for the different gauges, the percentage catch for each month was calculated using the monthly totals from the gauges mounted at 0.305 m (for PLU, TBR and STO) and 1.0 m (for PLU) and the reference gauge of each type mounted at 0.0 m. Figure 2 shows the clear evidence of undercatch, with the value in each month registering below the 100% reference rainfall line. Over the study period, the gauges mounted at 0.305 m recorded an average of 93.7% (0.3 m PLU), 94.4% (0.3 m TBR) and 94.7% (0.3 m STO) of the rainfall collected in the equivalent gauge installed at 0.0 m

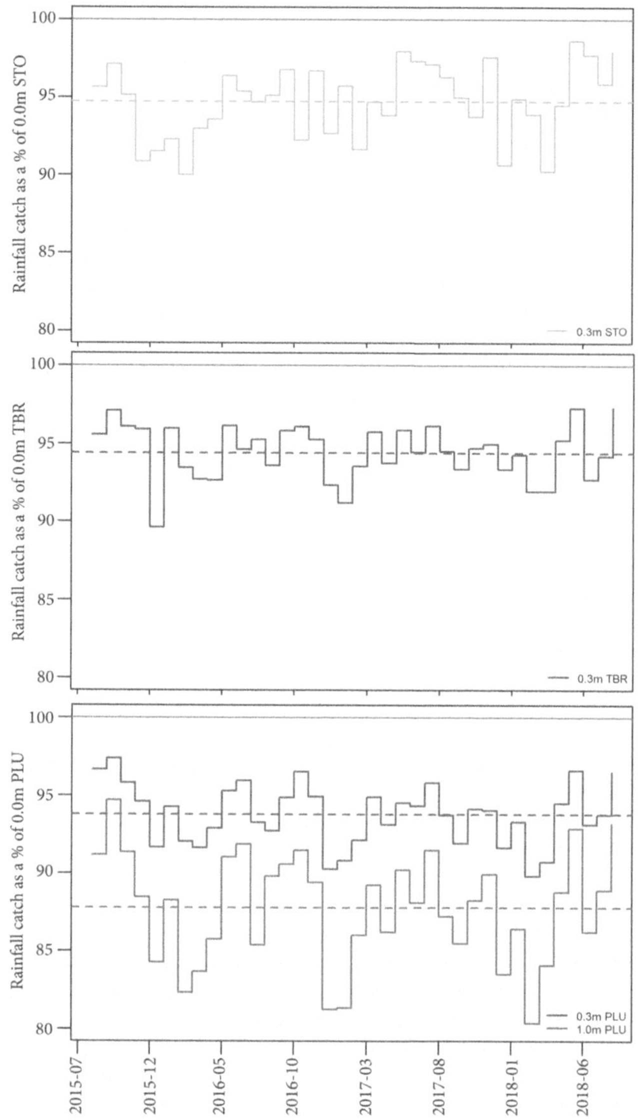

Figure 2 | Monthly undercatch in each raingauge, calculated as a percentage of the rainfall recorded in the reference gauge of that type (mounted at 0.0 m).

(i.e. a 5.3–6.3% undercatch). Significant variation was observed over the study period, ranging from a maximum monthly undercatch of 10.4% (December 2015, 0.3 m TBR) to a minimum of 1.4% (May 2018, 0.3 m STO). Notable variation was observed between the seasons (Figure 3), with the largest amounts of undercatch in the winter and smallest amounts in summer (0.3 m PLU, 0.3 m TBR)/autumn (0.3 m STO).

When compared with the 0.305 m gauges, increased amounts of undercatch were observed for the Ott Pluvio

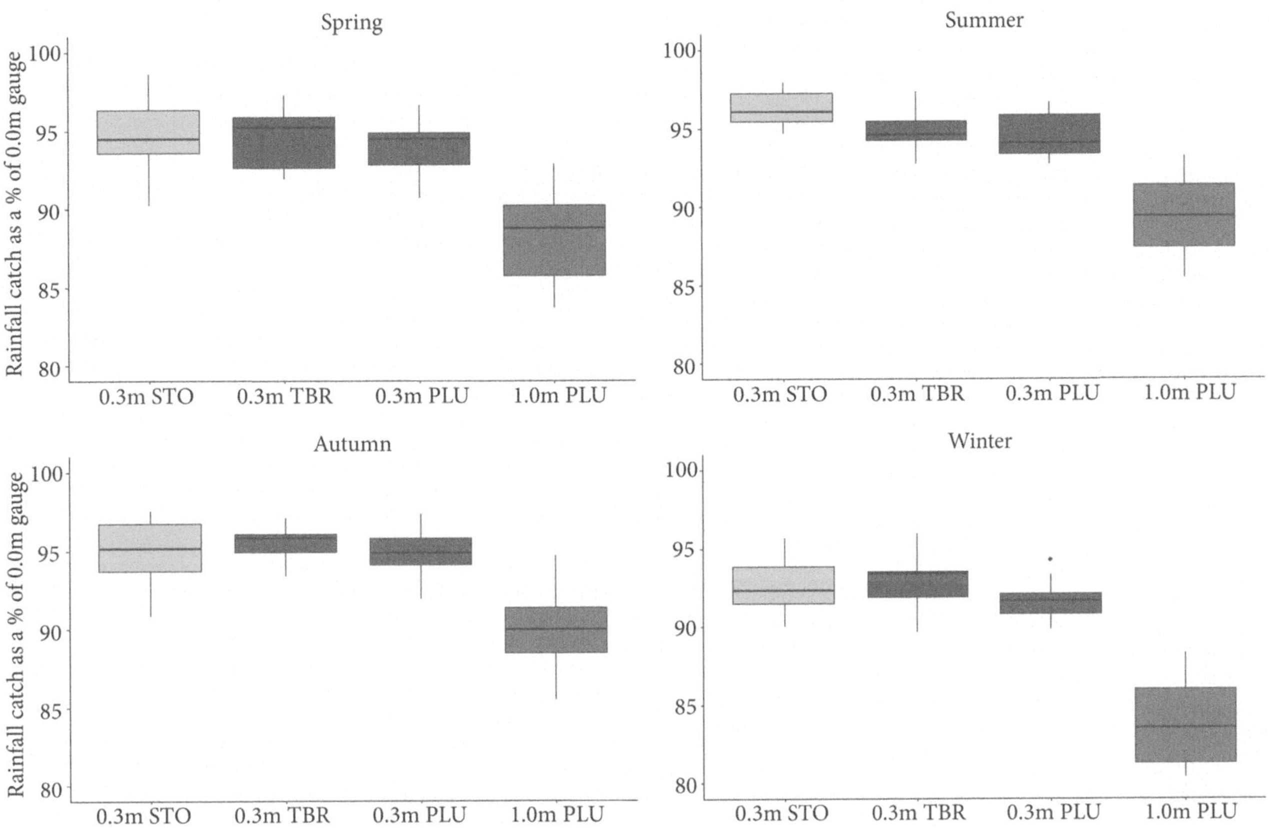

Figure 3 | Seasonal variation in undercatch recorded in each gauge as a percentage of the corresponding reference gauge.

with its rim mounted at a height of 1.0 m (Figure 2). Over the study period, the 1.0 m PLU recorded on average 87.3% of the rainfall recorded in the 0.0 m PLU, an undercatch of 12.7% – double that of the undercatch recorded in the 0.3 m PLU. There was also a wider range of values of monthly undercatch shown in the 1.0 m PLU – a maximum monthly undercatch of 19.6% (February 2018) and a minimum of 5.5% (September 2015). Similarly to the 0.305 m mounted gauges, the smallest amount of undercatch occurred in the summer and the largest in winter (Figure 3).

Event-based analysis

Using the dataset of 282 events, there is evidence of positive relationships between wind speed and undercatch for both Pluvio gauges (Figure 4(a)) which, while statistically significant ($p < 0.0001$), are weak with a large amount of scatter seen in the data. At low wind speeds (<1 m/s), the difference in undercatch between the gauges is smaller (although the 1.0 m PLU exhibits a wide range of undercatch during events) than at high wind speeds (>4 m/s), where there is a large difference in undercatch experienced between the gauges (Figure 4(b)).

A weak negative relationship was observed between undercatch and event average intensity (Figure 5(a)) but again, while statistically significant ($p < 0.001$), a large amount of scatter was seen, particularly at low intensities. The differences between the two gauges were more pronounced at lower intensities (Figure 5(b)); however, it should be noted that the dataset is highly skewed towards low-intensity events due to the climatology of the site (78% of the 282 events have an event average intensity of <2 mm/h) meaning that the effects of higher intensity events could not be well assessed.

The analysis presented in Figures 4 and 5 is based on the average intensity and wind speed of each event and when the event dataset was analysed using maximum data, a similar relationship was found. In order to explore the data further, four events were chosen for more

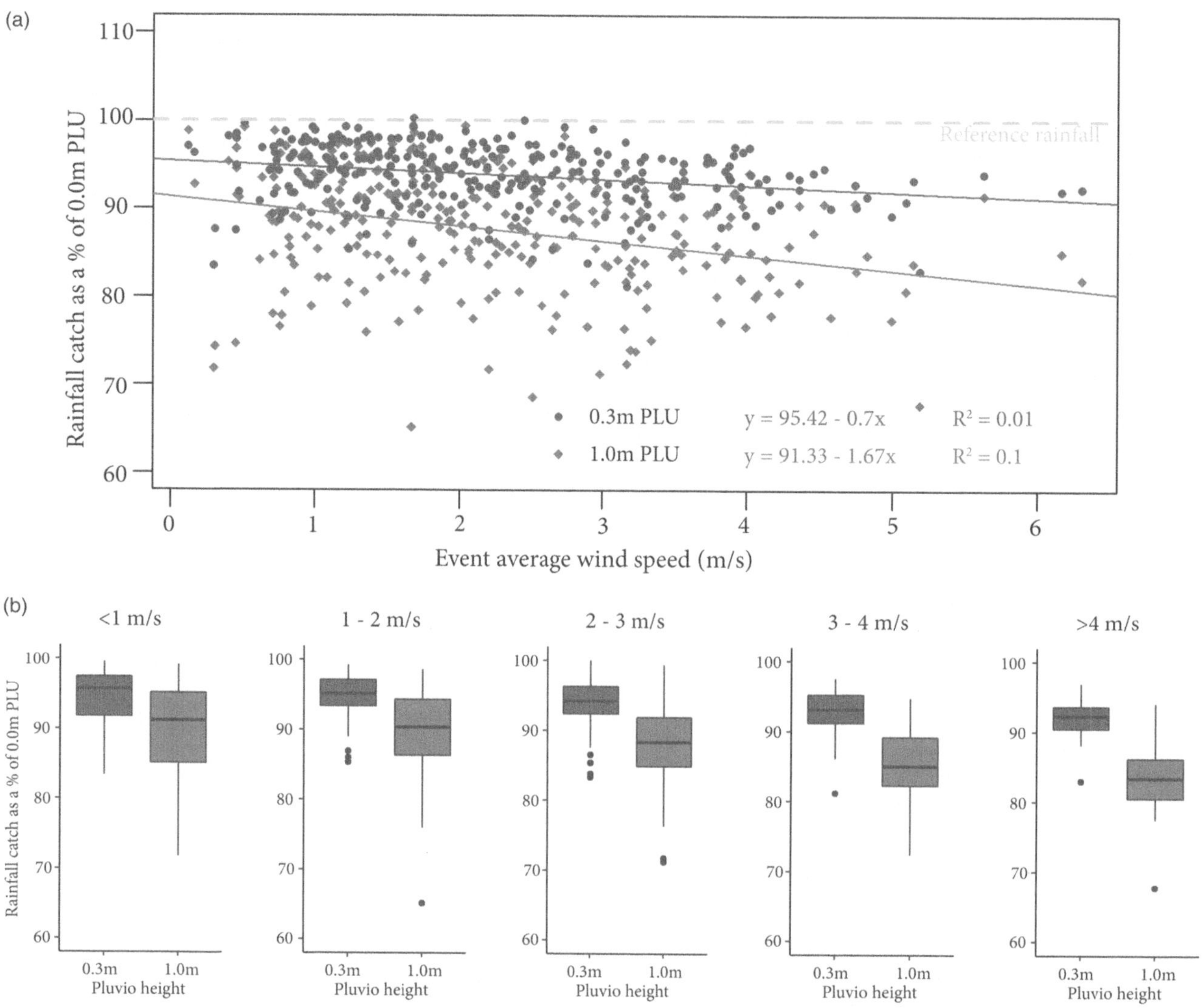

Figure 4 | Comparison between wind speed and undercatch recorded in 0.3 and 1.0 m PLU referenced to the 0.0 m PLU for all events (a) and subsets of wind speed (b).

detailed analysis. There was variation in undercatch observed during the events; however, there was no obvious pattern and, therefore, only the variation in wind speed and intensity are shown in Figure 6. The two low average intensity (<2 mm/h) events shown in Figure 6(a) and 6(b) exhibit large amounts of undercatch – 15–20% undercatch in the 0.3 m PLU and 25–30% undercatch in the 1.0 m PLU. In contrast, the two examples of high average intensity (>4 mm/h) events shown in Figure 6(c) and 6(d) exhibit small amounts of undercatch – 3–4% undercatch in the 0.3 m PLU and 5–8% undercatch in the 1.0 m PLU. Figure 6(d) shows the highest intensity event within the dataset (11 mm/h) which cause some localised surface water flooding in Wallingford. Within this event, it is evident that during the higher intensity period at the beginning where undercatch would be expected to be small (based on Figure 5), the concurrent high wind speed acts to increase the amount of undercatch (up to 8% in the 1.0 m PLU) and drops later on in the event (to only 1.5% in the 1.0 m PLU).

DISCUSSION

The results presented here for the three gauges mounted at 0.305 m compared with a reference pit gauge are

(a)

(b)

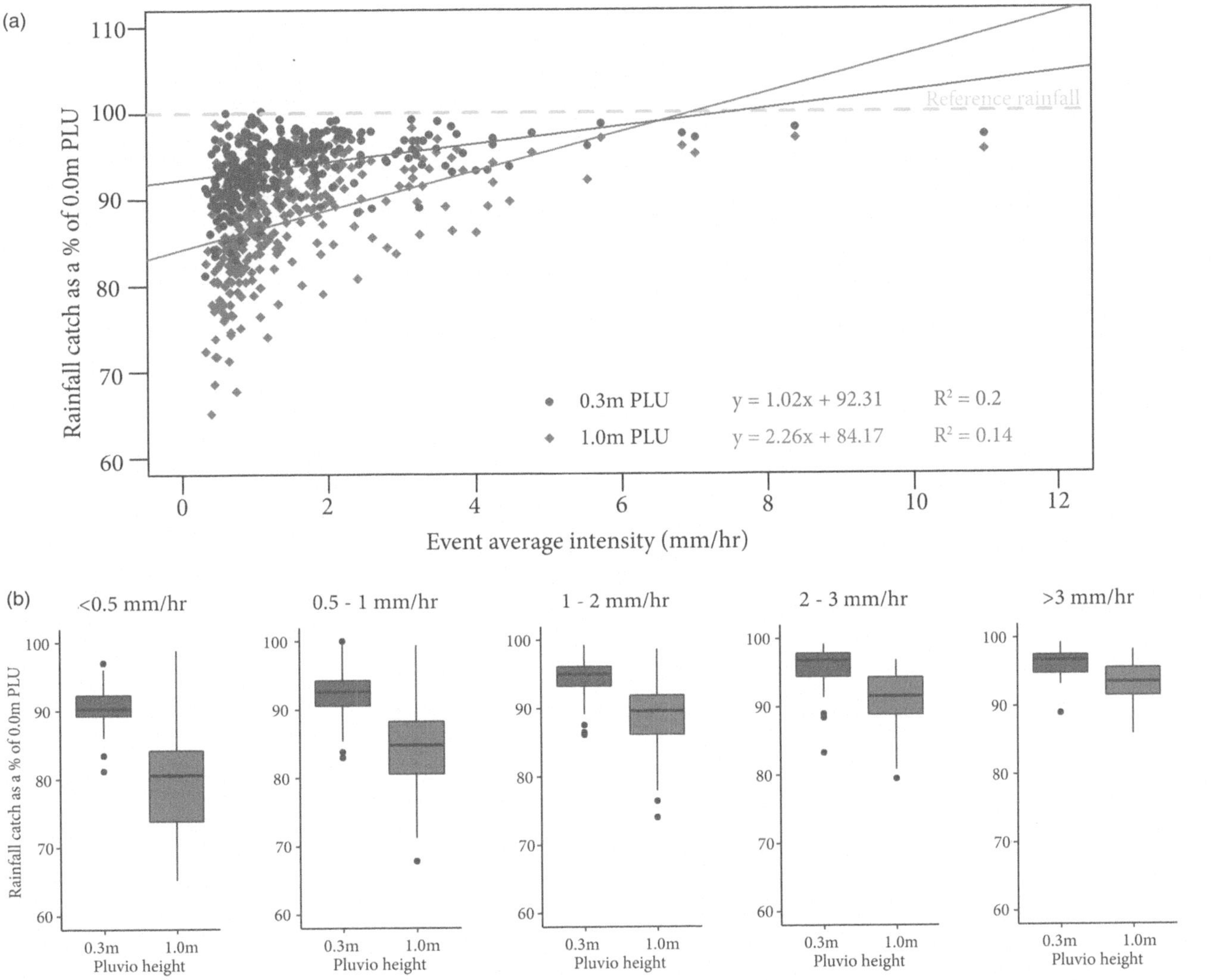

Figure 5 | Comparison between event average intensity and undercatch recorded in 0.3 and 1.0 m PLU referenced to the 0.0 m PLU for all events (a) and subsets of intensity (b).

comparable with previous studies at Wallingford published in Rodda (1968) and Rodda & Dixon (2012), which showed an average undercatch of 6.6% and 5.4%, respectively, compared with 5.3–6.3% reported here. The results here also agree with Rodda (1968) in terms of seasonality, with a maximum undercatch in winter and a minimum in summer, although there was variation between gauges. Similar values of undercatch have been presented in other studies within the UK, for example, Essery & Wilcock (1991) found an average of 5% undercatch between an exposed and pit storage gauge in Northern Ireland, although with temporal variation; and internationally, for example, Sieck *et al.* (2007) found 2–12% undercatch from gauges installed above the ground (at varying heights up to 0.5 m) in the USA.

All previous work published from Wallingford uses daily data from storage gauges, whereas this is the first presentation of data from automatic gauges of the type commonly installed across the current UK precipitation monitoring network. The broad similarity in undercatch between gauges of different types when sited at the UK standard 0.305 m is encouraging, suggesting that changing gauge types have little impact on wind-induced errors where the rim height is maintained at a constant. It should be noted however that, as stated in the Study Design section, the

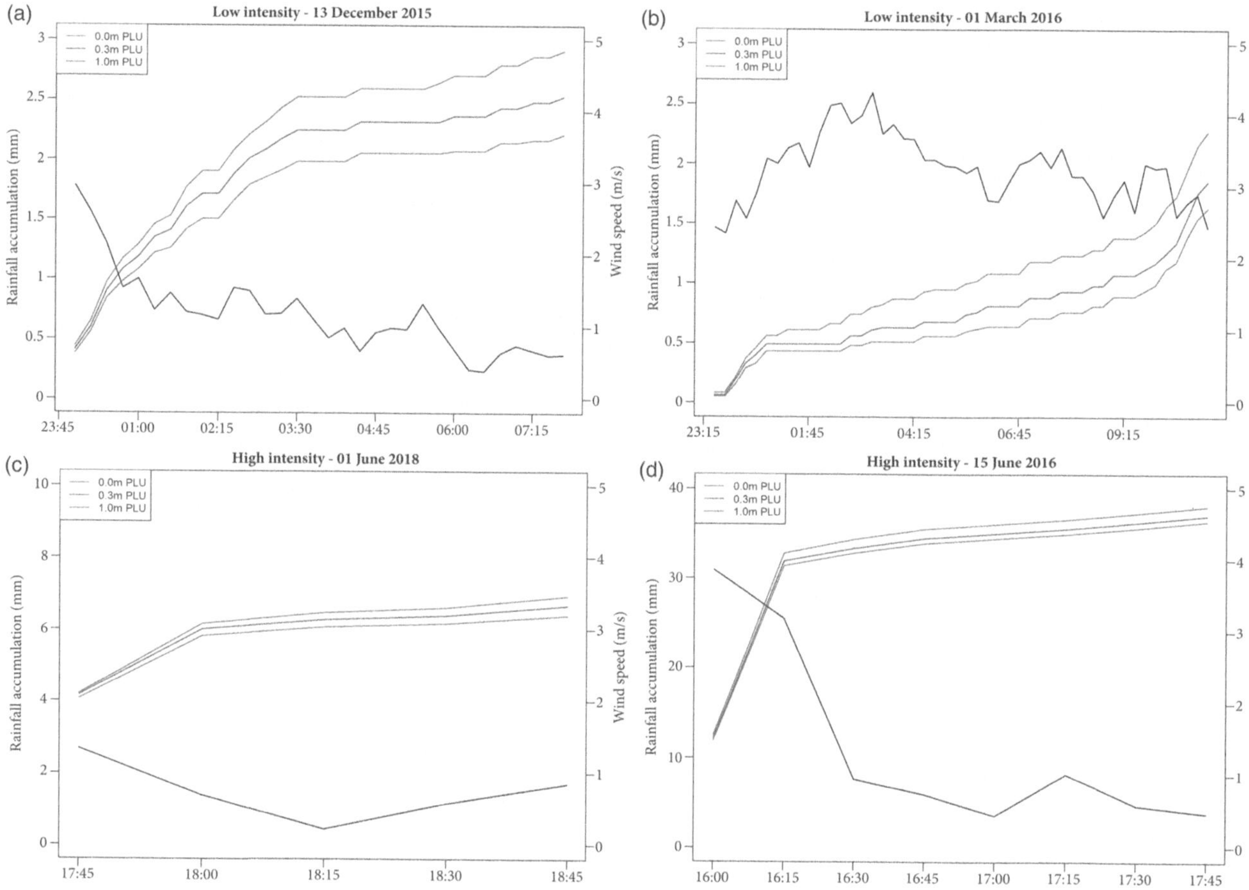

Figure 6 | Event rainfall accumulation and wind speed for two low event average intensity events ((a) and (b)) and two high event average intensity events ((c) and (d)).

effect of changing gauge type on other errors and uncertainties was isolated from this trial and thus this paper makes no comment on these.

The benefits of weighing gauges over TBRs have been well documented and have been a major driver in the installation of the gauges in the UK operational network. However, Pluvio gauges at a height of 1 m have generally replaced TBRs at 0.305 m, with the increase in height creating a larger disturbance of the surrounding airflow and installation in an area of higher wind speeds due to the boundary layer effect. In a computational fluid dynamics assessment, Colli *et al.* (2018) concluded 'the "chimney shaped" Ott Pluvio provides the least favourable aerodynamic performance when confronted by an airflow'. In this study, the installation of the Pluvio at the standard height of 1 m almost doubled the amount of undercatch on average compared with the same gauge installed at a height of 0.305 m (referenced to the pit gauge). The average of 12.7% (max 19.6%) undercatch for the Pluvio mounted at 1 m represents a substantial increase over that observed for lower gauges. The magnitude of the difference between the 0.305 and 1 m Pluvio gauges was generally more pronounced at times of higher wind speed and for lower intensity events. This finding is particularly notable given the lowland location of the Wallingford site. This percentage is comparable with gauges installed at a lower height at upland sites (e.g. 11.7% at a 0.5 m mounted gauge in Scotland (Pollock *et al.* 2018)) suggesting that even higher undercatch could be expected from 1 m gauges in upland locations. As weighing gauges could present the largest 'operational' benefits in upland areas where remote access means that reduced maintenance visits are a particular advantage (Grust & Stewart 2012), further investigation of undercatch at such sites is required.

The results presented here suggest that where the gauge rim height is maintained at a constant height, the change in gauge type from the daily storage gauges (which initially made up most of the UK observation network) to TBRs and more recently Pluvio WRGs does not significantly impact upon the wind-induced errors. This is perhaps in part because the shape of a Pluvio is not dissimilar to a conventional cylindrical-shaped TBR when mounted at 0.305 m (Figure 1). However, in the context of the UK network where a standard rim height of 0.305 m has been in use for over 100 years, the change to siting gauges with a rim height of 1 m may introduce significant inhomogeneity into rainfall records.

The lowering of rim heights on WRGs presents design challenges as they commonly need to incorporate a large collecting vessel to maximise the amount of rainfall they can catch between being emptied (either manually or via evaporation). One solution is, therefore, to install such gauges in pits, but it is acknowledged that the construction and maintenance of these for every gauge in the national network is unlikely to be practical. Therefore, in order to overcome this and yield usable data from a growing network of Pluvios in the UK operational network, correction factors need to be developed to adjust rainfall totals recorded at specific heights to equivalent rainfall totals recorded at the standard height – and preferably to ground level for purposes such as hydrology where the amount of precipitation reaching the ground is of interest. Previous studies involving the development of a correction factor have used linear regression (e.g. Essery & Wilcock 1991; Benning & Yang 2005) although only using relatively simple approaches applied at broad timescales. The relationships found in this study, although similar in direction to those of Seibert & Morén (1999) and Chvíla *et al.* (2005), are based on a short-length dataset with a large amount of scatter and a skewed intensity variable, therefore the development of correction factors is not recommended based on the data presented here.

For many applications, dynamic correction of data at a finer time-step would be essential as an investigation into the drivers of undercatch in this study shows marked variations within events (Figure 6). In order to further understand this relationship, higher temporal resolution wind speed and rainfall observations are needed. Anemometers installed at the rim height of each Pluvio with wind speed data delivered at a 1-min time resolution were not available during the study but have since been added to the next phase of the field trial to further develop an understanding of the relationship between wind speed and undercatch. The advantages of increased time resolution of rainfall measurements have been shown by Chvíla *et al.* (2005), but Nystuen *et al.* (1996) found increased noise in 1-min data from weighing gauges, which may make further investigations into undercatch more challenging.

Even with a more detailed understanding of the drivers of undercatch from higher resolution data, due to the variable nature of the UK climate (Parry *et al.* 2013), any correction factor that could be developed using the results of this study would be site-specific. An early attempt to understand this spatial variation in undercatch is presented in Rodda & Smith (1986) for the UK (1–16% undercatch) and Sevruk & Hamon (1984) internationally (0–30% undercatch). In a more recent study, Pollock *et al.* (2018) compared undercatch between a lowland site in England and an upland site in Scotland which varied between 3.4 and 11.2%, respectively, due to different wind velocities during events. It may be possible to develop spatially variable corrections based on site climatologies, although this is complicated by the lower density of the meteorological monitoring network in the UK compared with the raingauge network. Ultimately, however, a detailed understanding of the impact of wind-induced errors on rainfall observations across the UK is unlikely to be possible without the development of a network of reference gauges installed at ground level alongside gauges at other heights used in the network. Were such a capability to be established across the country, a national spatial correction grid could likely be developed and used to adjust both individual site records and/or in conjunction with existing areal rainfall products (e.g. CEH-GEAR; Keller *et al.* 2015).

CONCLUSION

This study presents, for the first time in the UK, a field trial of storage, TBRs and WRGs to investigate the impact of changes of raingauge siting within the UK operational network on wind-induced undercatch. These results presented here support the need to consider further the homogeneity of rainfall records in light of changes to gauge height within the UK, with the amount of undercatch almost double (on average) for the gauge installed at 1 m compared with 0.305 m. Investigations into the drivers of the observed undercatch using the event dataset suggest that the impact of gauge rim height is more evident at higher wind speeds and during low-intensity rainfall events but that the weakness of these relationships in the observed data prevented the development of a correction factor in this location.

The results of this study highlight the importance of understanding changes in raingauge undercatch as networks evolve. Where such changes are not communicated to data users and/or corrected for, they may complicate the detection of long-term changes in water balance calculation (e.g. Marsh & Dixon 2012) and the investigation of significant hydrological events (e.g. Barker *et al.* 2016). On the basis of this study, the authors aim to extend the field trials at Wallingford and use high-resolution rainfall and wind speed data to improve understanding and explore the generation of at-site correction factors. Improvements to the experimental set-up are also being considered, including options for measuring wind speed at the gauge rim height and dynamic calibration of the TRBs across a range of intensities (as recommended in CEN (2012)).

The results presented in this study are from a lowland UK site. Previous studies suggest that higher undercatch could be expected at more exposed upland sites and the knowledge of spatial variation in undercatch with varying topography is currently poorly understood, both in the UK and globally. As such, it is recommended that to prevent this introduction of inhomogeneity in time series, any future installation of replacement gauges are done at the UK standard height of 0.305 m or, to further improve the understanding of undercatch and develop adjustment methods for rainfall data, a network of International Standard reference pit gauges need to be constructed around the UK (and in other countries) with full metadata detailing their siting.

While the focus of this study has been on the inhomogeneity which may be introduced to the UK rainfall records by changing gauge rim height, the authors acknowledge that the UK practice of installing gauges at 0.305 m is different from that in many other countries. In locations where the standard installation of gauges is higher, it will be easier to maintain consistency with changing gauge designs. However, the magnitude of the undercatch observed in this study for gauges installed at 1 m is of relevance to all networks with rim heights of this level and above. Precipitation time series are often used to underpin a wide variety of hydrological science and operational water management decisions. For such applications, it is the amount of water reaching the ground surface which is often of interest and hence, unless properly understood and accounted for across space and time, undercatch of the magnitude of the 12.7% observed in this study and potentially even higher, represents a significant uncertainty in catchment hydrology and therefore the management of water resources and water-related hazards.

ACKNOWLEDGEMENTS

This work was supported by the Natural Environment Research Council award number NE/R016429/1 as part of the UK-SCaPE programme delivering National Capability. The authors thank Lucy Barker (CEH) for creating an R script to extract the rainfall events and to Lucy and Stephen Turner (CEH) for providing valuable comments while preparing the paper.

REFERENCES

Allerup, P. & Madsen, H. 1980 Accuracy of point precipitation measurements. *Nordic Hydrology* **11**, 57–70.

Barker, L., Muchan, K., Turner, S., Hannaford, J. & Parry, S. 2016 The winter 2015/2016 floods in the UK: a hydrological appraisal. *Weather* **71** (12), 324–333.

Benning, J. & Yang, D. 2005 Adjustment of daily precipitation data at Barrow and Nome Alaska for 1995–2001. *Arctic, Antarctic, and Alpine Research* **37** (3), 276–283. https://doi.org/10.1657/1523-0430(2005)037[0276:AODPDA]2.0.CO;2.

CEN 2010 *Hydrometry – Specification for a Reference Raingauge Pit, EN 13798:2010*. European Committee for Standardization (CEN), Brussels, Belgium.

CEN 2012 *Hydrometry – Measurement of the Rainfall Intensity (Liquid Precipitation): Requirements, Calibration Methods and Field Measurements, EN/TR 16469:2012*. European Committee for Standardization (CEN), Brussels, Belgium.

Chubb, T., Manton, M. J., Siems, S. T., Peace, A. D. & Bilish, S. P. 2015 Estimation of wind-induced losses from a precipitation gauge network in the Australian snowy mountains. *Journal of Hydrometeorology* **16** (6), 2619–2638. https://doi.org/10.1175/JHM-D-14-0216.1.

Chvíla, B., Sevruk, B. & Ondrás, M. 2005 The wind-induced loss of thunderstorm precipitation measurements. *Atmospheric Research* **77** (1–4), 29–38. https://doi.org/10.1016/j.atmosres.2004.11.032.

Ciach, G. J. 2003 Local random errors in tipping-bucket rain gauge measurements. *Journal of Atmospheric and Oceanic Technology* **20**, 752–759. https://doi.org/10.1175/1520-0426(2003)20<752:LREITB>2.0.CO;2.

Colli, M., Pollock, M., Stagnaro, M., Lanza, L. G., Dutton, M. & O'Connell, E. 2018 A computational fluid-dynamics assessment of the improved performance of aerodynamic rain gauges. *Water Resources Research* **54** (2), 779–796. https://doi.org/10.1002/2017WR020549.

Dunkerley, D. 2008 Identifying individual rain events from pluviograph records: a review with analysis of data from an Australian dryland site. *Hydrological Processes* **22**, 5024–5036. https://doi.org/10.1002/hyp.

Essery, C. I. & Wilcock, D. N. 1991 The variation in rainfall catch from standard UK meteorological office raingauges: a twelve year case study. *Hydrological Sciences Journal* **36** (1), 23–34. https://doi.org/10.1080/02626669109492482.

Gray, B. & Toucher, M. 2019 Rain gauge accuracy at a high-altitude meteorological station in Cathedral Peak. *Journal of Hydrologic Engineering* **24** (2). https://doi.org/10.1061/(ASCE)HE.1943-5584.0001741.

Grust, K. & Stewart, D. 2012 UK Trial of the OTT Pluvio. BHS Eleventh National Symposium, Hydrology for a Changing World, Dundee, pp. 01–07. https://doi.org/10.7558/bhs.2012.ns24.

Habib, E., Krajewski, W. F. & Kruger, A. 2001 Sampling errors of tipping-bucket raingauge measurements. *Journal of Hydrological Engineering* **6** (2), 159–166.

Keller, V. D. J., Tanguy, M., Prosdocimi, I., Terry, J. A., Hitt, O., Cole, S. J., Fry, M., Morris, D. G. & Dixon, H. 2015 CEH-GEAR: 1 km resolution daily and monthly areal rainfall estimates for the UK for hydrological and other applications. *Earth System Science Data* **7** (1), 143–155. https://doi.org/10.5194/essd-7-143-2015.

Kochendorfer, J., Rasmussen, R., Wolff, M., Baker, B., Hall, M. E., Meyers, T., Landolt, S., Jachcik, A., Isaksen, K., Braekkan, E. & Leeper, R. 2017 The quantification and correction of wind-induced precipitation measurement errors. *Hydrology and Earth System Sciences* **21**, 1973–1989. https://doi.org/10.5194/hess-21-1973-2017.

Lanza, L. G. & Vuerich, E. 2009 The WMO field intercomparison of rain intensity gauges. *Atmospheric Research* **94** (4), 534–543. https://doi.org/10.1016/j.atmosres.2009.06.012.

Looper, J. P., Vieux, B. E. & Moreno, M. A. 2012 Assessing the impacts of precipitation bias on distributed hydrologic model calibration and prediction accuracy. *Journal of Hydrology* **418–419**, 110–122. https://doi.org/10.1016/j.jhydrol.2009.09.048.

Marsh, T. & Dixon, H. 2012 *The UK Water Balance – How Much Has It Changed in a Warming World?* BHS Eleventh National Symposium, Hydrology for a Changing World, Dundee, UK, pp. 1–5.

Mekonnen, G. B., Matula, S., Doležal, F. & Fišák, J. 2015 Adjustment to rainfall measurement undercatch with a tipping-bucket rain gauge using ground-level manual gauges. *Meteorology and Atmospheric Physics* **127** (3), 241–256. https://doi.org/10.1007/s00703-014-0355-z.

Michaelides, S., Levizzani, V., Anagnostou, E., Bauer, P., Kasparis, T. & Lane, J. E. 2009 Precipitation: measurement, remote sensing, climatology and modeling. *Atmospheric Research* **94** (4), 512–533. https://doi.org/10.1016/j.atmosres.2009.08.017.

Molini, A., Lanza, L. G. & La Barbera, P. 2005 Improving the accuracy of tipping-bucket rain records using disaggregation techniques. *Atmospheric Research* **77** (1–4), 203–217. https://doi.org/10.1016/j.atmosres.2004.12.013.

Nespor, V. & Sevruk, B. 1999 Estimation of wind-induced error of rainfall gauge measurements using a numerical simulation. *Journal of Atmospheric and Oceanic Technology* **16**, 450–464.

Nystuen, J. A., Proni, J. R., Black, P. G. & Wilkerson, J. C. 1996 A comparison of automatic rain gauges. *Journal of Atmospheric and Oceanic Technology* **13**, 62–73. https://doi.org/10.1175/1520-0426(1996)013<0062:ACOARG>2.0.CO;2.

Parry, S., Marsh, T. & Kendon, M. 2013 2012: from drought to floods in England and Wales. *Weather* **68** (10), 268–274.

Pollock, M. D., O'Donnell, G., Quinn, P., Dutton, M., Black, A., Wilkinson, M. E., Colli, M., Stagnaro, M., Lanza, L. G., Lewis, E., Kilsby, C. G. & O'Connell, P. E. 2018 Quantifying and mitigating wind-induced undercatch in rainfall measurements. *Water Resources Research* **54** (6), 3863–3875. https://doi.org/10.1029/2017WR022421.

Rodda, J. C. 1968 The rainfall measurement problem. *Proceedings IASH General Assembly*, Berne, Switzerland, 1967. IAHS/Publ. No 78, pp. 21.

Rodda, J. C. & Dixon, H. 2012 Rainfall measurement revisited. *Weather* **67** (5), 131–136. https://doi.org/10.1002/wea.1884.

Rodda, J. C. & Smith, S. W. 1986 The significance of the systematic error in rainfall measurement for assessing wet deposition. *Atmospheric Environment* **20**, 1059–1064.

Savina, M., Schäppi, B., Molnar, P., Burlando, P. & Sevruk, B. 2012 Comparison of a tipping-bucket and electronic weighing precipitation gauge for snowfall. *Atmospheric Research* **103**, 45–51. https://doi.org/10.1016/j.atmosres.2011.06.010.

Seibert, J. & Morén, A. S. 1999 Reducing systematic errors in rainfall measurements using a new type of gauge. *Agricultural and Forest Meteorology* **98–99**, 341–348. https://doi.org/10.1016/S0168-1923(99)00107-0.

Sevruk, B. 1996 Adjustment of tipping-bucket precipitation gauge measurements. *Atmospheric Research* **42** (1–4), 237–246. https://doi.org/10.1016/0169-8095(95)00066-6.

Sevruk, B. & Chvíla, B. 2005 Error sources of precipitation measurements using electronic weight systems. *Atmospheric Research* **77** (1–4), 39–47. https://doi.org/10.1016/j.atmosres.2004.10.026.

Sevruk, B. & Hamon, W. 1984 *International Comparison of National Precipitation Gauges with a Reference Pit Gauge.* World Meteorological Organization, Geneva, Switzerland.

Sevruk, B., Ondrás, M. & Chvíla, B. 2009 The WMO precipitation measurement intercomparisons. *Atmospheric Research* **92** (3), 376–380. https://doi.org/10.1016/j.atmosres.2009.01.016.

Sieck, L. C., Burges, S. J. & Steiner, M. 2007 Challenges in obtaining reliable measurements of point rainfall. *Water Resources Research* **43** (1), 1–23. https://doi.org/10.1029/2005WR004519.

Stisen, S., Hojberg, A. L., Troldborg, L., Refsgaard, J. C., Christensen, B. S. B., Olsen, M. & Henriksen, H. J. 2012 On the importance of appropriate precipitation gauge catch correction for hydrological modelling at mid to high latitudes. *Hydrology and Earth System Sciences* **16** (11), 4157–4176. https://doi.org/10.5194/hess-16-4157-2012.

Strangeways, I. 2004 Improving precipitation measurement. *International Journal of Climatology* **24** (11), 1443–1460. https://doi.org/10.1002/joc.1075.

Strangeways, I. 2010 A history of rain gauges. *Weather* **65** (11), 311. https://doi.org/10.1002/wea.548.

Tapiador, F. J., Turk, F. J., Petersen, W., Hou, A. Y., García-Ortega, E., Machado, L. A. T., Angelis, C. F., Salio, P., Kidd, C., Huffman, G. J. & de Castro, M. 2012 Global precipitation measurement: methods, datasets and applications. *Atmospheric Research* **104–105**, 70–97. https://doi.org/10.1016/j.atmosres.2011.10.021.

Tian, X., Dai, A., Yang, D. & Xie, Z. 2007 Effects of precipitation-bias corrections on surface hydrology over northern latitudes. *Journal of Geophysical Research Atmospheres* **112** (14). https://doi.org/10.1029/2007JD008420.

Upton, G. J. G. & Rahimi, A. R. 2003 On-line detection of errors in tipping-bucket raingauges. *Journal of Hydrology* **278** (1–4), 197–212. https://doi.org/10.1016/S0022-1694(03)00142-2.

WMO 2008 *Guide to Hydrological Practices.* Volume I: Hydrology – From Measurement to Hydrological Information. https://doi.org/10.1080/02626667.2011.546602

First received 21 February 2019; accepted in revised form 6 October 2019. Available online 25 November 2019

Hydrological modelling of climate change impacts on river flows in Siberia's Lena River Basin and implications for the Atlantic Meridional Overturning Circulation

C. E. Hudson and J. R. Thompson

ABSTRACT

A hydrological model of Siberia's Lena River Basin is calibrated and validated against observed river discharge at five stations. Implications of the Representative Concentration Pathway 4.5 scenario for river discharge are assessed using projections from 41 Coupled Model Intercomparison Project Phase 5 General Circulation Models grouped into 12 genealogical-based groups as well as a group ensemble mean. Annual precipitation increases in all scenarios (1.7–47.4%). Increases in annual PET are of a similar range (6.0–45.5%). PET peaks in June compared to July for the baseline. All temperature changes exceed 1.5 °C (range: 2.2 °C–6.2 °C). The largest absolute increases are in winter (maximum +7 °C). Changes in mean annual discharge range from −8.5 to +69.9%. Ten GCM groups and the group ensemble mean project increases. Earlier snowmelt is dominant so the annual flood peaks in May compared with June for the baseline. Increased discharge of the Lena and other Eurasian rivers to the Arctic Ocean has the potential to impact Atlantic Meridional Overturning Circulation (AMOC). Enhanced fluxes for four groups are capable of weakening the AMOC. Changes for other groups may contribute to weakening when combined with other sources of freshwater and warmer temperatures.

Key words | Atlantic Meridional Overturning Circulation (AMOC), climate change, Coupled Model Intercomparison Project Phase 5 (CMIP5), Lena, RCP4.5

C. E. Hudson (corresponding author)
J. R. Thompson
UCL Department of Geography,
University College London,
Gower Street, London WC1E 6BT,
UK
E-mail: *charlotte.hudson.15@ucl.ac.uk*

INTRODUCTION

Climate change will intensify the global hydrological cycle. Modified precipitation patterns coupled with changes in temperature and evapotranspiration will have important implications for river discharge (Vihma *et al.* 2016). The most severe hydrometeorological impacts of rising temperatures are being observed in, and are projected for, the Arctic, with mean annual air temperatures between 2001 and 2012 being 1.5 °C warmer than during 1971–2000 (Overland *et al.* 2013). Precipitation is increasing and is projected to be >50% higher by 2100. Winter warming is projected to be four times greater than summer warming, modifying snowmelt, evapotranspiration and ultimately river discharge (Ye *et al.* 2004).

Reported increases in Arctic river flows have raised concerns about the integrity of Atlantic Meridional Overturning Circulation (AMOC; Shu *et al.* 2017). AMOC comprises northward flows of warm saline water, the formation of North Atlantic Deep Water (NADW) through sinking due to buoyancy loss and southward return flows of cold deepwater (Buckley & Marshall 2016). Palaeoclimate proxy records (e.g. Broecker *et al.* 1985; Clark *et al.* 2002) suggest that the AMOC has collapsed in the past pointing to the potential for it having stable 'on' and 'off' states. Simulations from simple numerical models (e.g. Manabe & Stouffer 1988; Hawkins *et al.* 2011) support the presence of this bi-stable behaviour suggesting that it may collapse in the future. Enhanced freshwater input into the Arctic Ocean could reduce surface water density and potentially inhibit the formation of deep water, causing positive feedback whereby reduced NADW formation decreases northward transport

doi: 10.2166/nh.2019.151

of saltwater further reducing water density and, therefore, convection. However, studies using coupled general circulation models (GCMs) to assess AMOC alterations have not identified this instability, and the most recent consensus of the Intergovernmental Panel on Climate Change is that AMOC slowdown is more likely than complete collapse during the 21st century (Kirtman *et al.* 2013).

The implications of a weakening or collapse of AMOC would be widespread due to global-scale teleconnections (Vellinga & Wood 2008). Climatic implications may include North Atlantic cooling, an equatorward shift of the Inter-Tropical Convergence Zone and weakened monsoons (Buckley & Marshall 2016). AMOC collapse may also increase water resources stress in Europe and southern Asia due to altered precipitation patterns (Gosling 2013). Additionally, reductions in the extent of boreal and temperate forests are projected, with implications for carbon storage in these latitudes (Köhler *et al.* 2005). Fisheries and crop yields could be negatively impacted due to changes in ocean circulation with the potential for major societal implications (Keller *et al.* 2000; Kuhlbrodt *et al.* 2009). Given the significance of changes in AMOC, this study investigates the potential impacts of climate change upon river flows within Siberia's Lena River Basin, a major contributor of freshwater to the Arctic. Results are scaled up to assess the potential implications for the AMOC of changes in Eurasian runoff to the Arctic Ocean.

STUDY AREA: THE LENA RIVER BASIN

The Lena River (Figure 1), which enters the Arctic Ocean via the Laptev Sea, is located in northern Asia and

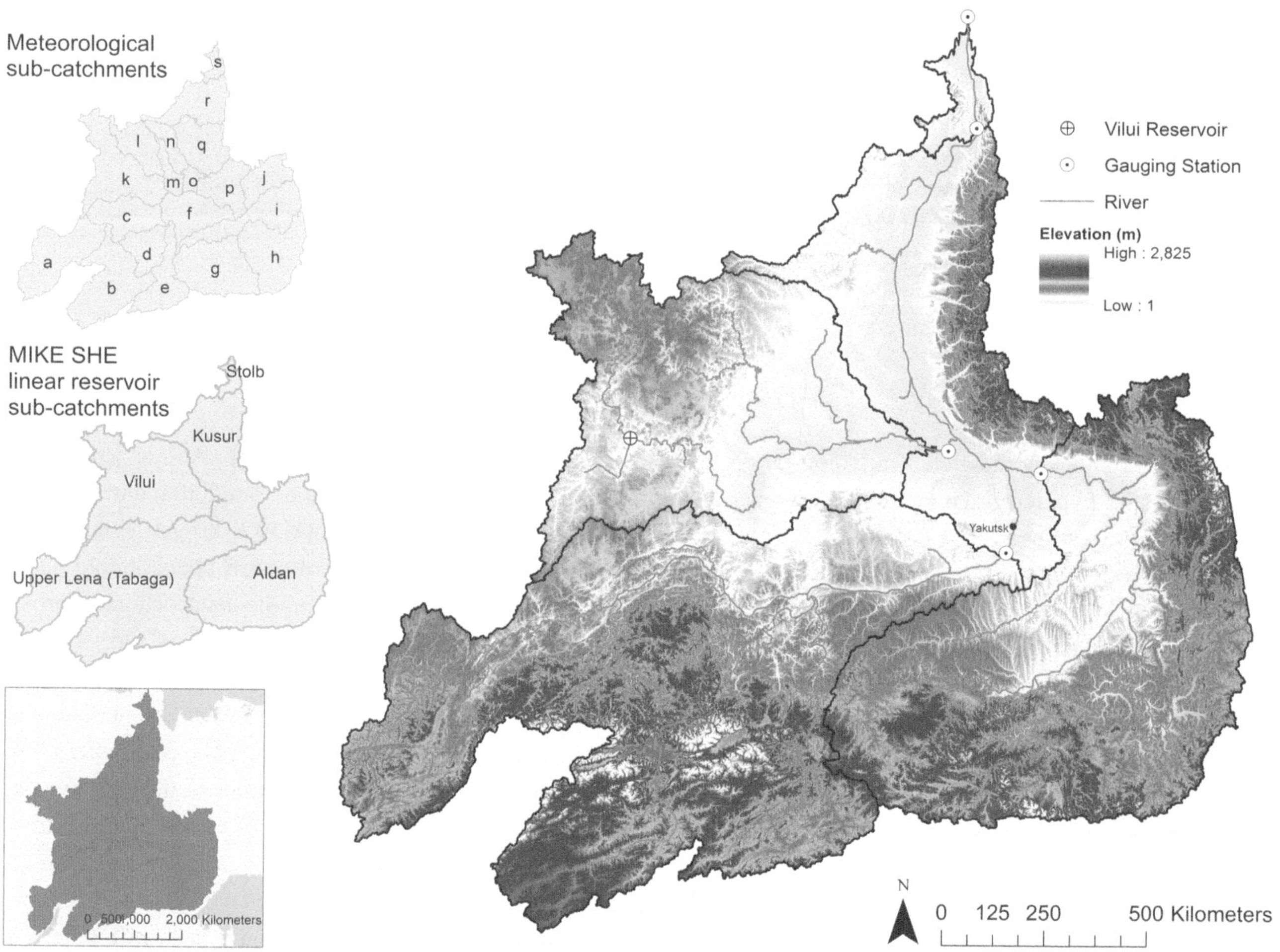

Figure 1 | The Lena River Basin including the locations of five gauging stations for which river discharge is simulated and the sub-catchments used within the MIKE SHE model.

originates in the Baikal mountains (maximum altitude: 1,640 m). It is the eleventh longest river in the world (4,400 km) with the ninth largest basin (32,000 km^2) (Gelfan *et al.* 2017). As the second largest Eurasian river in terms of discharge, following the Yenisei and preceding the Ob, the Lena provides around 15% of the total mean annual runoff to the Arctic Ocean (mean annual discharge: 524 km^3; Shiklomanov *et al.* 2000; Ye *et al.* 2004) although it varies from year to year.

The Lena Basin lies in a zone of continental moderate and sub-arctic climate (Liu & Yang 2011). Precipitation is highest during April–October (total precipitation at Yakutsk = 152 mm), peaking in July and subsequently decreasing during November–March (total precipitation at Yakutsk = 78 mm). Mean annual precipitation (based on CRU TS4.01) varies from 402 mm over the Tabaga sub-catchment to 280 mm over Stolb. This downstream decline is repeated for temperature with the mean annual temperature (CRU TS4.01) decreasing from −8 °C over Tabaga to −17 °C over Stolb. Temperature and evapotranspiration peak in July, after which snow accumulation commences, reaching a maximum extent in November, before snowmelt begins in March (Ye *et al.* 2003).

The lowest and highest river flows occur during winter and summer, respectively. Snowmelt during May causes rapid increases in discharge, which on average peaks in June (Gelfan *et al.* 2017). Permafrost underlays 93% of the basin and directs precipitation and snowmelt to rivers. It contributes to low sub-surface storage capacity, causing large differences between winter and summer flows (Ye *et al.* 2004). The Lena Basin has three main tributaries, the Aldan, Upper Lena and Vilui (Figure 1). The Aldan experiences peak flows that are approximately 60 times the lowest flows in April. Relatively higher high- and low-flows are experienced in the Upper Lena so that the ratio of the highest to lowest flows is 26 (Ye *et al.* 2003). The Vilui contributes a relatively small amount to annual runoff (9% of discharge). The reservoir on this tributary (completed in 1967) has a capacity equivalent to 7% of annual runoff. Whilst it increases winter flows above natural levels (Ye *et al.* 2003), these account for just 10% of annual discharge, and the higher summer flows are relatively unaffected (Holmes *et al.* 2012).

The basin is sparsely populated and vegetation is largely natural comprising forests (84%), shrublands (9%), grasslands (3%), croplands (2%) and wetlands (1%) (Liu & Yang 2011). Forests dominate the southwest and tundra dominates the north. Whilst the basin's water resources are utilised for domestic purposes, hydropower and irrigation, the total use comprises a very low percentage of the mean annual runoff (Berezovskaya *et al.* 2005). The impacts of climate change can, therefore, be more easily identified and are likely to dominate future changes as opposed to anthropogenic activities.

Temperature and precipitation have increased across the basin, especially during the cold season (November–April; Dzhamalov *et al.* 2012). Changes in river discharge include earlier seasonal peaks and larger flows in spring, summer and winter in contrast to autumn declines. A significant upward trend (up to 90%) at the basin outlet (Stolb) during low-flow periods has been recorded, whilst slight increases (5–10%) in high-flows have been observed (Ye *et al.* 2003). Recognition that climate change may have already impacted the discharge of the Lena River, combined with the important role these flows and those from the other Eurasian rivers play in the global climate system, provides the impetus for assessing potential future changes within the current study.

METHODS

Model development, calibration and validation

This study employs a coupled hydrological/hydraulic model of the Lena River Basin developed using the MIKE SHE/MIKE 11 modelling system. MIKE SHE is commonly described as a deterministic, fully distributed and physically based hydrological modelling system, although it includes a range of process descriptions, some of which are more conceptual and semi-distributed in nature (Refsgaard *et al.* 2010). MIKE SHE is dynamically coupled to MIKE 11, a 1D hydraulic model that represents the channel flow (e.g. Thompson *et al.* 2004). Model development for the Lena Basin followed approaches used in other large river systems (e.g. Andersen *et al.* 2001; Thompson *et al.* 2013). Table 1 summarises the model set-up and the data it employs. Basin extent was determined using the USGS GTOPO-30 DEM with the lowest point defined as Stolb. A cell size of

Table 1 | Data used in the MIKE SHE/MIKE 11 model of the Lena River Basin

Component within model	Data source and derivation	Use within the MIKE SHE/MIKE 11 model
Basin area	Obtained using the USGS GTOPO30 DEM[a]	Defined the model domain within MIKE SHE and specified as a shapefile.
Topography	Values extracted from the USGS GTOPO-30 DEM	Defined the topography within MIKE SHE. Specified as a grid file.
Sub-catchments	The USGS GTOPO-30 DEM, R-ArcticNET[b] gauging station locations and major tributaries.	Defined five main sub-catchments within MIKE SHE. Specified as a shapefile. Also defined the linear reservoir sub-catchments and baseflow sub-catchments. A grid file defined the locations of the smaller 19 meteorological sub-catchments, which allowed the spatial distribution of climatic variables. Within these areas, parameters were adjusted during model calibration.
Land use	USGS 1 km Global Land Cover Characterisation data[c]	Defined the spatial distribution of land use within the Lena River Basin. Twenty-nine original classes were re-classified into nine classes. Specified as a grid file.
Vegetation properties: RD and LAI	Values from the literature (Arnell 2005)	Root depth was defined for each land-cover class. This value describes the depth of the zone from evapotranspiration can occur. These were constant for each land-cover classes. LAI describes the ratio of the leaf area to the ground area.
River network	Using the USGS GTOPO-30, the river network was identified using ArcMap Hydrology Tools.	A shapefile was specified in MIKE 11. It was then manually digitised to define the river network.
Cross-sections	Identified and measured using Google Earth Pro. Elevations were extracted from the basin DEM.	Defined channel cross-sections within MIKE 11. Each channel width was assigned a stream order. Elevations were assigned to each cross-section.
Overland Flow: Manning Number	Values from the literature using the approach of Thompson *et al.* (2013).	This was spatially distributed throughout the catchment based on the overlying vegetation. Specified as a grid file. Defined the rate at which overland flow is routed to channels.
Unsaturated Zone: Soil properties	FAO Digital Soil Map of the World.[d] Using ArcMap, the basin was separated into three main soil classes. Values from the literature (Atwell *et al.* 1999).	Defined for the water content at saturation, water content at field capacity, water content at the wilting point and saturated hydraulic conductivity within each soil class.

[a]lta.cr.usgs.gov/GTOPO30.
[b]www.r-arcticnet.sr.unh.edu/v4.0/index.html.
[c]lta.cr.usgs.gov/GLCC.
[d]www.fao.org/soils-portal/soil-survey/soil-maps-and-databases.

10 km × 10 km (total cells: 24,680) was employed with GTOPO-30 being used to define the elevation of each grid cell. The saturated zone was represented using linear reservoirs, a conceptual, semi-distributed approach, particularly applicable to large river systems where the focus is river flow simulation (Andersen *et al.* 2001). The Lena Basin was divided into five sub-catchments (defining the extent of a series of saturated zone linear reservoirs – see Figure 1) based on the GTOPO-30 DEM and the location of gauging stations used for model calibration and validation. Stations were selected based on the length and completeness of discharge records within the Regional Arctic Hydrographic Network (R-ArcticNET) dataset. Each sub-catchment was further sub-divided according to elevation into zones of approximately equal size (14 in total) representing the highest, intermediate and lowest zones. These were specified as interflow reservoirs, whilst two baseflow reservoirs representing faster and slower baseflow storage were specified

beneath each sub-catchment. The two time constants (interflow and percolation) for each interflow reservoir and the baseflow time constant for each baseflow reservoir, which control exchanges between reservoirs and the MIKE 11 hydraulic model, were varied during calibration.

The unsaturated zone was simulated using the two-layer water balance method. Spatial distribution of soil types was based on the FAO Digital Soil Map of the World (v3.6, 2003) with groups aggregated into three textural classes: 'fine', 'medium/fine' and 'coarse'. Hydraulic parameters were taken from Atwell *et al.* (1999). Land cover was based on the USGS Global Land Cover Characterisation dataset (GLCC). The dominant classes, including deciduous needle-leaf forest, bare rock and tundra, were retained. The remaining classes were aggregated into groups of similar characteristics, including water, croplands and grasslands, broadleaf forest, needle-leaf evergreen shrubs and bogs. For each, root depth (RD) and Leaf Area Index (LAI) were obtained from Arnell (2005). Permafrost was not included in the model as this is not a feature of MIKE SHE. This is common to a number of other hydrological models that have been used to assess the impacts of climate on river flows including within high latitude basins such as the Lena Basin (e.g. Gosling *et al.* 2017; Veldkamp *et al.* 2018).

To account for variations in climate, the areas defining the extent of the five saturated zone linear reservoirs were further divided into a total of 19 smaller areas herein referred to as meteorological sub-catchments. The discretisation of these areas was based on their ranges in latitude, longitude and elevation, as well as the major tributaries within each saturated zone linear reservoir sub-catchment (Figure 1). Time-series of mean monthly precipitation and monthly maximum and minimum temperatures were derived for each meteorological sub-catchment from the CRU TS4.01 dataset (Harris *et al.* 2014). Since R-ArcticNET data used in model calibration and validation comprised a mean monthly discharge necessitating the aggregation of simulated mean daily discharge for comparison, precipitation was distributed evenly and temperatures assumed constant on a daily basis through each month. Whilst this is acknowledged to be a simple approach, it follows earlier work undertaken using MIKE SHE and other hydrological models in similarly large river basins that demonstrated insensitivity to alternative temporal disaggregation of meteorological data when simulation results are aggregated to mean monthly discharges (e.g. Kingston *et al.* 2011; Thompson *et al.* 2013). Following approaches in other mountainous settings, varying precipitation lapse rates were applied over these sub-catchments (Ji & Luo 2013) and were subject to calibration but kept within the bounds used elsewhere (Immerzeel *et al.* 2012; Thompson *et al.* 2014). CRU TS4.01 temperatures for each meteorological sub-catchment were used to calculate the Hargreaves potential evapotranspiration (PET; Hargreaves & Samani 1985). MIKE SHE then calculates actual evapotranspiration using the evaporative demand (PET), crop coefficients and the available soil moisture. PET, as for precipitation, was evenly distributed through each month on a daily basis. This PET method is recommended as an alternative to Penman–Monteith in cases of limited data availability (Allen *et al.* 1998) and has been used in similar studies (e.g. Ho *et al.* 2016; Thompson *et al.* 2017a). Snowmelt was simulated using a degree-day method and meteorological sub-catchment averaged CRU TS4.01 temperature. As for precipitation, temperature lapse rates were specified within each meteorological sub-catchment.

A digitised channel network defined the MIKE 11 river branches comprising the main river channels. All branches were specified as coupled to MIKE SHE. Cross-sections were established using channel widths obtained from Google Earth and estimated maximum depths based on similar studies (Thompson *et al.* 2014). The Vilui reservoir was excluded from the section of the MIKE 11 model within this sub-catchment due to a lack of data and its small influence on the annual discharge (Holmes *et al.* 2012).

R-ArcticNET discharge data were separated into two periods, 1960–1979 and 1980–1999, for calibration and validation, respectively. In both cases, the previous year was used as a spin-up period. Calibration was undertaken from upstream to downstream by adjusting the parameters defined above (principally the saturated zone linear reservoir time constants and lapse rates). As previously indicated, simulated river discharges, which were stored at the maximum model time step of 24 hours, were aggregated to mean monthly discharge for comparison with the R-ArcticNET data. Model performance was assessed visually and statistically using the Nash–Sutcliffe efficiency coefficient

(NSE), bias (*Dv*) and the Pearson correlation coefficient (*r*). Performance based on the values of these three statistics was further classified into one of five classes (ranging between 'very poor' and 'excellent') using the scheme of Ho *et al.* (2016), which was itself adapted from Henriksen *et al.* (2003).

Climate change scenarios

Precipitation and minimum and maximum temperatures were obtained for the 41 GCMs of Phase 5 of the Climate Model Intercomparison Project (CMIP5) and the Representative Concentration Pathway (RCP) 4.5 scenario as it represents the most likely increase in global temperatures (UNFCCC 2015). The use of an ensemble of climate models enables assessment of the magnitude of GCM-related uncertainty (Ho *et al.* 2016). Using the mean output from a range of GCMs to force a hydrological model is thought to provide a more reliable representation of future conditions than the output from a single GCM. However, this assumption only holds if the GCMs are independent of one another (Pirtle *et al.* 2010). This is not strictly the case for the CMIP5 ensemble since GCMs developed by different institutions share literature, parameter values and some model code, and the ensemble includes multiple versions of some GCMs or numerous GCMs from a single institution. The potential for biases due to this lack of model independence was addressed by grouping the 41 GCMs according to their genealogy using 12 groups (Ho *et al.* 2016; Table 2).

Mean monthly maximum, mean and minimum temperatures and precipitation were obtained for each meteorological sub-catchment for the baseline (1961–1990) and scenario (2071–2100) periods for all 41 GCMs. This 30-year scenario period was selected to represent conditions towards the end of the 21st century (e.g. Thompson *et al.* 2017b). The baseline is of identical length and incorporates most of the period used in model calibration/validation, but excludes the latter part of the 20th century during which changes in meteorological networks may impact model performance (discussed below). The mean values were then obtained from the GCMs in each of the 12 groups. Monthly differences (°C for temperature, % for precipitation) between baseline and scenario meteorological conditions were calculated for each GCM group for each of the meteorological sub-catchments. These differences, referred to as delta factors, were subsequently used to perturb the original CRU TS4.01 precipitation and temperature data, and then Hargreaves PET was re-calculated. The delta factor approach ensures that scenario time-series retain the baseline climate variability and are not affected by any biases inherent within an individual GCM (Anandhi *et al.* 2011). An additional group ensemble mean scenario was established using the same approach and employing the mean monthly baseline and scenario temperatures and precipitation from the 12 groups.

RESULTS

Model calibration and validation

Figure 2 demonstrates the generally good model performance. Timings of low- and high-flows are well represented, with slightly earlier increases in simulated discharges at Kusur and Stolb. Annual peaks are well reproduced upstream. Although the model is less successful at simulating low-flows at Vilui towards the end of the period (most likely due to the dam), the rising and recession limbs are well represented. This generally superior upstream performance is further demonstrated by the observed and simulated river regimes (mean monthly discharge) for each gauging station and the calibration, validation and baseline periods.

NSE for the calibration period is classified as 'excellent' at two stations and 'very good' at the remaining three stations (Table 3). Lower NSE values at Kusur and Stolb are related to the poorer representation of peak discharges. It was not possible to increase peaks without impacting the annual rise and recession, and ultimately increasing the overall bias. Since the focus of this study is the volume of water flowing into the Arctic Ocean, calibration was focussed on achieving a good match between the observed and simulated mean flows. The bias for one station (Tabaga) was classified as 'excellent', whilst for the remaining stations it was 'very good'. The values of *r* were varied, as calibration was a compromise between achieving higher *r* values and smaller biases.

NSE values for the validation period are 'excellent' or 'very good' at all stations, except Vilui where winter flows

Table 2 | The CMIP5 GCMs and their grouping by genealogy

No	GCM	Institution	GCM group
1	ACCESS1.0	Commonwealth Scientific and Industrial Research Organisation (CSIRO) and Bureau of Meteorology (BOM), Australia	10
2	ACCESS1.3		10
3	BCC-CSM1.1	Beijing Climate Center, China Meteorological Administration	12
4	BCC-CSM1.1(m)		12
5	BNU-ESM	College of Global Change and Earth System Science, Beijing Normal University	12
6	CanESM2	Canadian Centre for Climate Modelling and Analysis	1
7	CCSM4	National Center for Atmospheric Research	12
8	CESM1(BGC)	Community Earth System Model Contributors	12
9	CESM1(CAM5)		12
10	CMCC-CM	Centro Euro-Mediterraneo per I Cambiamenti Climatici	11
11	CMCC-CMS		11
12	CNRM-CM5	Centre National de Recherches Météorologiques/Centre Européen de Recherche et Formation Avancée en Calcul Scientifique	11
13	CSIRO-Mk3.6.0	Commonwealth Scientific and Industrial Research Organisation in collaboration with the Queensland Climate Change Centre of Excellence	2
14	EC-EARTH	EC-Earth consortium	11
15	FGOALS-g2	LASG, Institute of Atmospheric Physics, Chinese Academy of Sciences	3
16	FIO-ESM	The First Institute of Oceanography, SOA, China	12
17	GFDL-CM3	NOAA Geophysical Fluid Dynamics Laboratory	6
18	GFDL-ESM2G		6
19	GFDL-ESM2M		6
20	GISS-E2-H p1	NASA Goddard Institute for Space Studies	7
21	GISS-E2-H p2		7
22	GISS-E2-H p3		7
23	GISS-E2-H-CC		7
24	GISS-E2-R p1		7
25	GISS-E2-R p2		7
26	GISS-E2-R p3		7
27	GISS-E2-R-CC		7
28	HadGEM2-AO	Met Office Hadley Centre (additional HadGEM2-ES realizations contributed by Instituto Nacional de Pesquisas Espaciais)	10
29	HadGEM2-CC		10
30	HadGEM2-ES		10
31	INM-CM4	Institute for Numerical Mathematics	4
32	IPSL-CM5A-LR	Institut Pierre-Simon Laplace	8
33	IPSL-CM5A-MR		8
34	IPSL-CM5B-LR		8
35	MIROC5	Atmosphere and Ocean Research Institute (the University of Tokyo), National Institute for Environmental Studies, and Japan Agency for Marine-Earth Science and Technology	9

(*continued*)

Table 2 | continued

No	GCM	Institution	GCM group
36	MIROC-ESM	Japan Agency for Marine-Earth Science and Technology, Atmosphere and Ocean Research Institute (the University of Tokyo), and National Institute for Environmental Studies	9
37	MIROC-ESM-CHEM		9
38	MPI-ESM-LR	Max-Planck-Institut für Meteorologie (Max Planck Institute for Meteorology)	11
39	MPI-ESM-MR	Meteorological Research Institute	11
40	MRI-CGCM3		5
41	NorESM1-M	Norwegian Climate Centre	12

Genealogy-based GCM groups

No	Group name	Number of GCMs	No	Group name	Number of GCMs	No	Group name	Number of GCMs
1	CanESM2	1	5	MRI-CGCM3	1	9	MIROC	3
2	CSIRO-Mk3.6.0	1	6	GFDL	3	10	UKMO	5
3	FGOALS-g2	1	7	GISS	8	11	European	6
4	INM-CM4	1	8	IPSL	3	12	NCAR	8

are underestimated. A shift from overestimation, or small underestimation, during the calibration period, to underestimation during the validation period is evident. *Dv* values ranged from 'good' at Tabaga, Aldan and Stolb, to 'poor' at Kusur and Vilui. This poorer performance may be related to changes in meteorological networks and how well they represent Lena's climate. These factors (discussed below) may have been particularly acute towards the end of the 20th century. If so, they are less likely to impact the baseline period against which climate change results are compared. NSE for this period is classified as 'very good' for three stations and 'good' for two stations (Table 3). *Dv* is classified as 'excellent' at four stations and 'very good' at the remaining station. Figure 2 confirms the generally very good performance of the model for this period.

Projected climate

Mean annual precipitation, temperature and PET are projected to increase for all GCM groups across the Lena Basin (Figure 3). The magnitude of these changes varies between the groups and sub-catchments. Seasonal patterns of change are also variable, most prominently at higher latitudes; where some groups (2, 6, 8, 9, 10 and 11) project a second precipitation peak in October in addition to the July baseline peak. In general, but with the exception of Group 5, larger increases in precipitation are projected downstream. The largest increase in annual precipitation across all groups and sub-catchments is 47.4% (Group 9, sub-catchment *r*), whilst the smallest is 1.7% (Group 4, sub-catchment *d*). Group 10 is associated with the largest inter-sub-catchment range (11.9–45.1%) and Group 7 the smallest (12.5–20.7%). Changes for the group ensemble mean range between 15.0 and 27.5% (mean: 19.7%).

All temperature increases exceed the 2015 Paris target of 1.5 °C (UNFCCC 2015) varying between 2.2 °C and 6.2 °C (mean: 2.7 °C). The greatest absolute increases are projected during winter (maximum 6.2 °C, Group 9, sub-catchment *r*). The duration of the period when temperatures are above freezing extends by, on average, 1 month, most prominently at higher latitudes. Group 9 is associated with the largest increases (mean: 5.4 °C), including earlier seasonal gains in temperature. Group 10 again has the largest inter-sub-catchment range of change (2.0 °C). In contrast, groups 4 and 5 project relatively small increases (2.2 °C–3.2 °C and 2.2 °C–3.5 °C, respectively). Increases in temperature for the group ensemble mean range between 3.2 °C and 4.4 °C (mean: 3.7 °C).

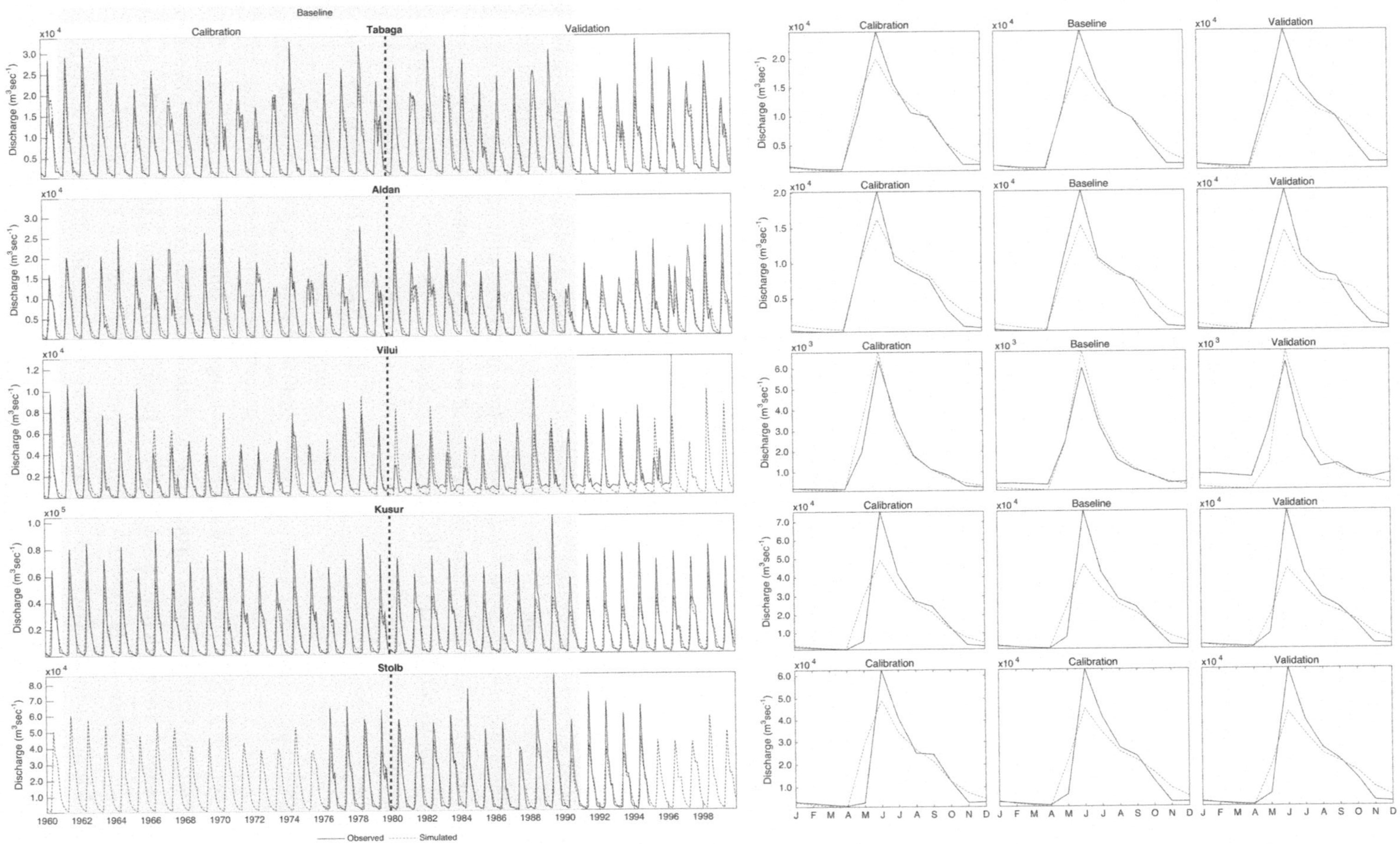

Figure 2 | Observed and simulated mean monthly discharge and river regimes at the five gauging stations within the Lena River Basin for the calibration (1960–1979), baseline (1961–1990) and validation (1980–1999) periods. Note different *y*-axis scales.

Table 3 | Model performance statistics and classification for the calibration (Cal, 1960–1979), baseline (1961–1990) and validation (Val, 1980–1999) periods

Station	Period	*Dv*		NSE		*r*	
Tabaga	Cal	−3.27	✓✓✓✓✓	0.90	✓✓✓✓✓	0.95	✓✓✓✓✓
	Baseline	−3.87	✓✓✓✓✓	0.83	✓✓✓✓	0.95	✓✓✓✓✓
	Val	−10.00	✓✓✓	0.83	✓✓✓✓	0.93	✓✓✓✓
Aldan	Cal	8.14	✓✓✓✓	0.88	✓✓✓✓✓	0.95	✓✓✓✓✓
	Baseline	3.23	✓✓✓✓✓	0.83	✓✓✓✓	0.95	✓✓✓✓✓
	Val	−8.25	✓✓✓✓	0.79	✓✓✓✓	0.92	✓✓✓✓
Vilui	Cal	5.33	✓✓✓✓	0.82	✓✓✓✓	0.90	✓✓✓✓
	Baseline	−2.17	✓✓✓✓✓	0.66	✓✓✓✓	0.87	✓✓✓
	Val	−16.47	✓✓	0.49	✓✓	0.77	✓
Kusur	Cal	−5.03	✓✓✓✓	0.74	✓✓✓✓	0.87	✓✓✓
	Baseline	−7.87	✓✓✓✓	0.63	✓✓✓	0.88	✓✓✓
	Val	−12.70	✓✓✓	0.76	✓✓✓✓	0.92	✓✓✓✓
Stolb	Cal	5.69	✓✓✓✓	0.70	✓✓✓✓	0.83	✓✓
	Baseline	2.44	✓✓✓✓✓	0.61	✓✓✓	0.76	✓
	Val	−5.22	✓✓✓✓	0.80	✓✓✓✓	0.92	✓✓✓✓
Performance indicator		**Excellent** ✓✓✓✓✓	**Very good** ✓✓✓✓	**Fair** ✓✓✓	**Poor** ✓✓		**Very poor** ✓
Dv		<5%	5–10%	10–20%	20–40%		>40%
NSE		>0.85	0.65–0.85	0.50–0.65	0.20–0.50		<0.20
r		≥0.95	0.90–0.94	0.85–0.89	0.80–0.84		<0.80

Note: Model performance classification is based on Ho *et al.* (2016) and adapted from Henriksen *et al.* (2003).

Increases in mean annual PET are of a similar range, albeit slightly smaller, to those of precipitation (6.0–45.5% across all GCM groups and sub-catchments). The smallest increases are predominantly projected by Group 5 (6.0–15.2%, mean: 10.9%), whilst Group 1 generally produces the largest increases (24.7–34.5%, mean: 27.8%). The range for the group ensemble mean is 15.5–24.2% (mean: 19.2%). All groups and the group ensemble mean project basin-wide peaks in June, 1 month earlier than for the baseline (although the largest absolute changes occur in May; Figure 3).

Projected river discharge

Changes in discharge are generally consistent with 10 of the 12 groups and the group ensemble mean projecting increases at all gauging stations. These increases are, however, of variable magnitude (Figure 4). Across the basin changes range between −8.5 and +36.8%. Groups 1, 3 and 5 project the largest basin-wide increases. Declines are limited to groups 4 and 12, which project declines at four (−8.5 to − 1.0%) and five (−5.8 to − 1.7%) stations, respectively. These groups are associated with relatively large increases in PET (8.0–19.0% and 18.1–26.2%, respectively) that exceed increases in precipitation (1.7–17.2% and 9.6–19.5%, respectively). The group ensemble mean projects increases in the mean discharge of between 5.6 and 18.6% (mean: 10.1%) with the increase of 9.2% for Stolb, indicative of Arctic Ocean inflow, contrasting with the range for the 12 groups of −5.3 to 21.7%. All but two groups (again 4 and 12) are associated with increases in these flows.

High (*Q*5) and low (*Q*95) flows also increase for most groups. Changes in *Q*5 across all groups and gauging stations range between −2.8 and +69.9%. Increases are, in percentage terms, larger than those for the mean annual discharge. Declines are again limited to groups 4 (three stations) and 12 (one station). However, they are small compared with most increases. The group ensemble mean projects increase in *Q*5 at all stations (range: 10.2–30.2%). *Q*95 increases in most cases with relatively small (≤6.7%) declines limited to just two stations for Group 4 and one for groups 2 and 12. The small (2.8%) decline for Tabaga

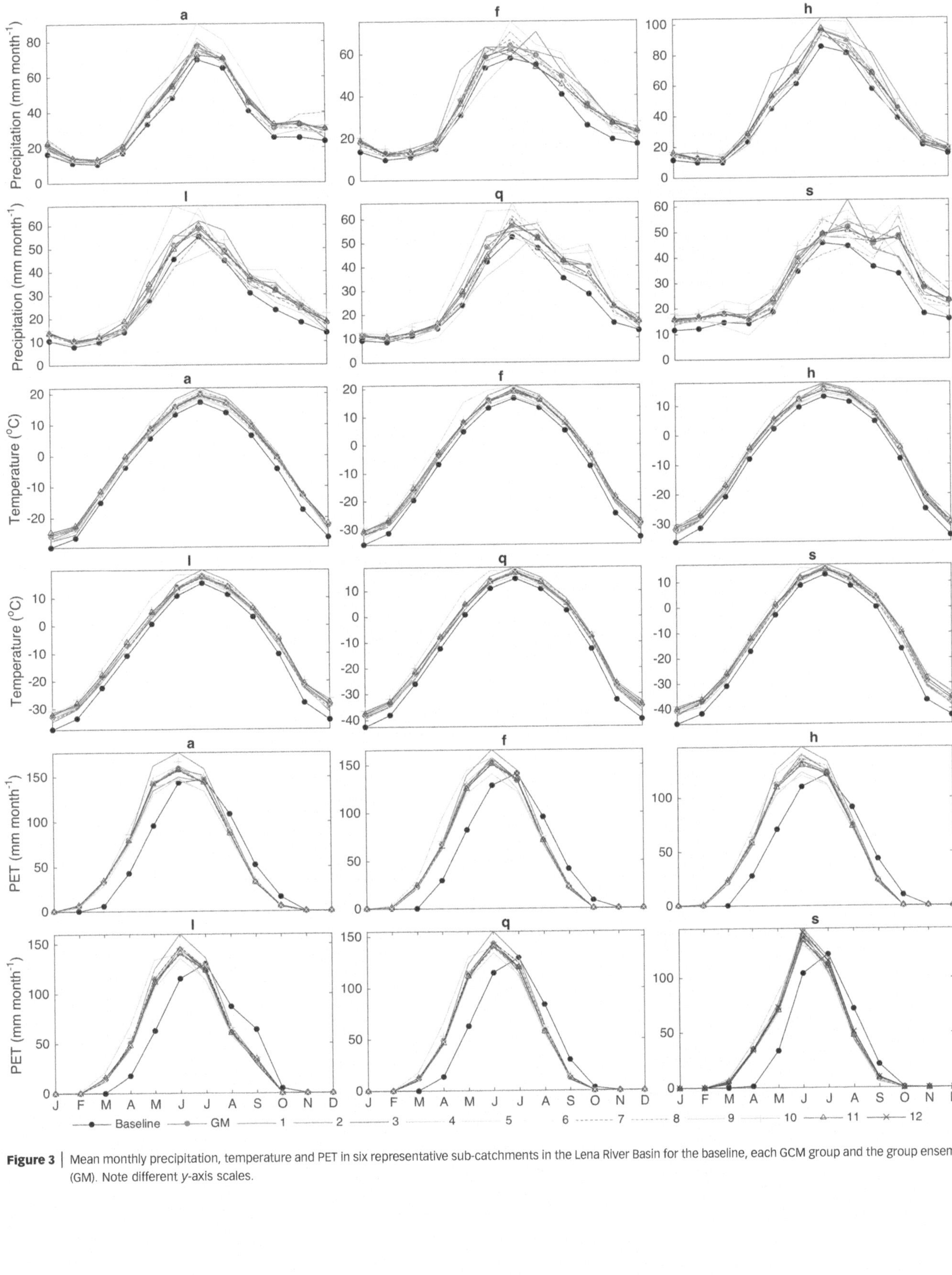

Figure 3 | Mean monthly precipitation, temperature and PET in six representative sub-catchments in the Lena River Basin for the baseline, each GCM group and the group ensemble mean (GM). Note different *y*-axis scales.

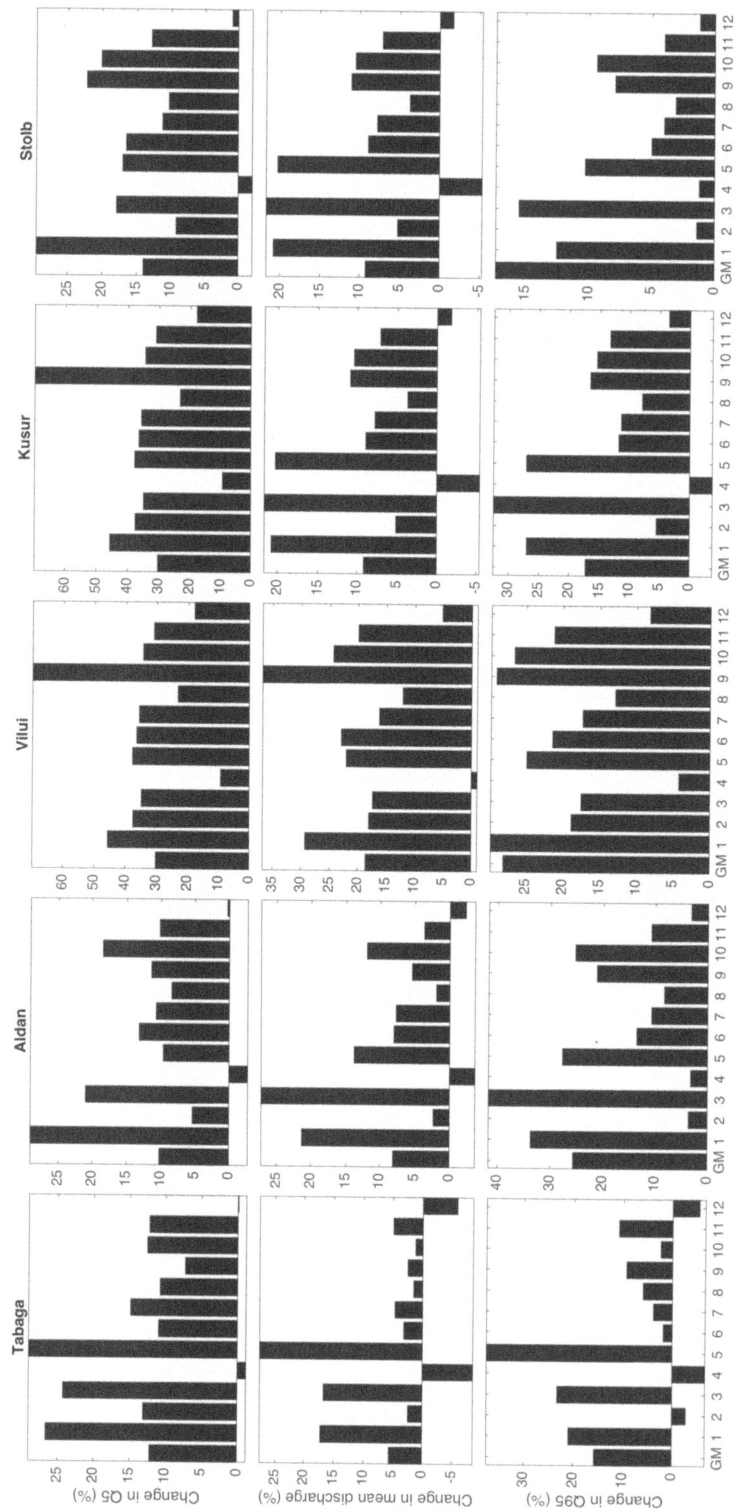

Figure 4 | Percentage changes in mean discharge, high (Q5) and low (Q95) discharges for each gauging station in the Lena River Basin. Note different *y*-axis scales.

projected by Group 2 is the only reduction in any discharge measure beyond groups 4 and 12. These two groups project the smallest increases in *Q*95 (<8.2%), whilst groups 1, 3 and 5 project some of the largest (up to 41.7%, Group 3, Aldan). Increases in *Q*95 of between 15.7 and 28.0% are projected by the group ensemble mean.

Projected river regimes (Figure 5) show that in many cases the seasonal peak advances to May compared with June under baseline conditions. This is most pronounced for groups 9 and 10, both of which project large basin-wide increases in temperature, and Group 11 and the group ensemble mean at Vilui. Group 9 projects the most pronounced change at Stolb with the mean May discharge being 82% larger than the baseline. For many groups, the recession limb declines more rapidly so that discharges in September are lower than during baseline conditions. The largest reductions at Stolb (19.0%) are projected by Group 4.

DISCUSSION

Model performance

This study expands research into the impacts of climate change on river discharge within the Arctic (e.g. Peterson *et al.* 2002; Arnell 2005), including, in comparison to other studies of the Lena Basin (Ye *et al.* 2003), extending the geographical range downstream to Stolb.

Model performance for the baseline period was classified as at least 'very good', and in some cases 'excellent'. It was comparable to, and in some instances better than, other models of the Lena and similar basins (e.g. Gosling *et al.* 2017; Veldkamp *et al.* 2018). Model performance was relatively weaker at Vilui possibly due to a lack of information regarding the reservoir and thus its exclusion from the model. Similar issues have been experienced elsewhere (Ho *et al.* 2016). Whilst performance for the validation period based on NSE was, in most cases, at least 'very good', discharges were notably underestimated, especially downstream. As previously stated, it is possible that this relates to a decline in how well the data used to force the model represents the basin's climate. Gridded CRU TS4.01 data are produced through interpolation of observations from meteorological stations (Harris *et al.* 2014). However, Arctic climatic observations are fraught with uncertainties due to sparse station networks, biases in measurements and changes in measurement methods (Rawlins *et al.* 2006). Within the Russian Arctic, both the instruments used to measure precipitation and the frequency of observations have changed over time. There is also potential for underestimation of precipitation due to difficulties in measuring snow, especially during windy winter conditions (Groisman *et al.* 1991). This could explain underestimated discharges at the most northern stations since unrealistically low winter snowfall will limit the volume of simulated spring meltwater. A mismatch between increasing Arctic discharge and declining or plateaued precipitation has been attributed to the closure of multiple meteorological stations in the late 20th century (Groisman *et al.* 1991). Many high elevation stations were lost and thus, interpolation is based on the stations at lower elevations with potentially less precipitation (Wang *et al.* 2016). Poor model performance for the latter validation period supports the argument that these problems were particularly acute towards the end of the last century.

Projected hydrometeorological changes within the Lena Basin

Results suggest relatively small inter-GCM variability in projected temperatures across the Lena Basin. Inter-GCM variability is larger for precipitation and PET with the range of change being slightly greater for precipitation, replicating results from other similar studies (Ho *et al.* 2016; Thompson *et al.* 2017a).

In general, model results suggest that discharge of the Lena River and its main tributaries will increase. This echoes findings of other studies that have highlighted increasing Arctic river flows (Peterson *et al.* 2002) and those that project future increases (Arnell 2005; Gosling *et al.* 2011; 2017; Hattermann *et al.* 2017). The shift towards earlier snowmelt floods has been reported in Siberia and throughout the Arctic (e.g. Overeem & Syvitski 2010; Vihma *et al.* 2016). Projected increases in winter precipitation, and hence deeper snow pack, will also contribute to higher spring discharges (Ye *et al.* 2004). The dominance of steeper recessions following the annual peak replicates results from Woo *et al.* (2008) that were attributed to increases in PET in excess of gains in precipitation.

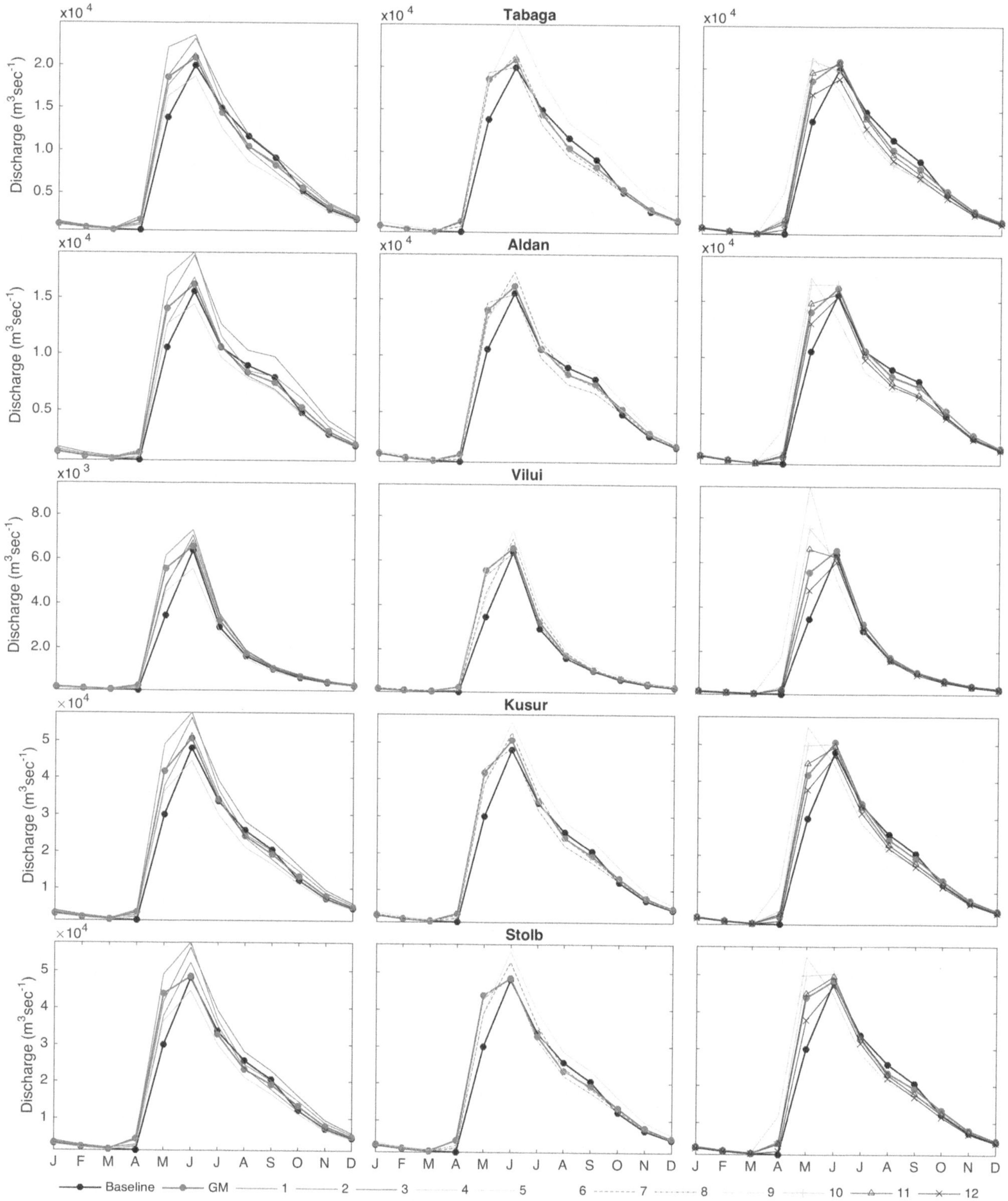

Figure 5 | Mean monthly discharges at each gauging station for each scenario. The baseline and group ensemble mean (GM) are shown in each figure to facilitate comparison. Note different *y*-axis scales for different stations.

Although increased discharges dominate scenario results, the changes vary considerably in magnitude. This uncertainty could be constrained using GCM weightings (e.g. Maxino *et al.* 2008) following the approach of Krysanova *et al.* (2018) who recommend assessing models based on their performance, then weighting or excluding them as appropriate. This could reduce the number of GCM groups, and therefore, the number of scenarios and subsequent uncertainty. Nonetheless, utilising this approach introduces questions regarding model exclusion, weight complexity and their derivation (Zaherpour *et al.* 2019), which potentially adds further uncertainty.

In common with similar studies (e.g. Arnell 2005; Thompson *et al.* 2013), potential changes in vegetation or anthropogenic interventions were not explicitly considered. Vegetation will shift northward with altered climate regimes (Vihma *et al.* 2016) with potential hydrological feedbacks (Arnell 2005). The omission of such features will have hydrological implications for processes such as PET, interception and infiltration. The incorporation of future land-use projections, such as those used within the Inter-Sectoral Impact Model Intercomparison Project (ISI-MIP2b; Frieler *et al.* 2017), which themselves are impacted by climatic and socio-economic drivers, would enable these impacts to be simulated. Permafrost melt, which as previously noted was not included in this and other models used to assess climate change impacts on the Lena and other similar basins (e.g. Gosling *et al.* 2017; Veldkamp *et al.* 2018), would also have hydrological implications including enhanced infiltration that may contribute to increasing groundwater contributions to river flow (Walvoord & Striegl 2007; Vihma *et al.* 2016). In contrast, continued melting may enhance vertical flow paths enough that increased infiltration reduces the volume of water reaching the river (Walvoord & Kurylyk 2016). These changes may be most significant in higher latitude sub-catchments due to the greater projected temperature increases (Gautier *et al.* 2018).

This study focussed on GCM-related uncertainty in future hydrometeorological conditions within the Lena Basin and did not consider hydrological model-related uncertainty. This could be investigated by simulating the same climate change scenarios with a number of hydrological models of the Lena using alternative model codes to MIKE SHE, the different process descriptions available within MIKE SHE or alternative parameterisations and spatial distributions of model input data (e.g. Thompson *et al.* 2013, 2014; Robinson 2018). Whilst the overall fraction of uncertainty within hydrological impact studies of climate change that is attributable to different hydrological models has been shown to be smaller than that due to different GCMs (e.g. Krysanova *et al.* 2017), choice of hydrological model may not be insignificant where processes are implemented uniquely in different models (Hattermann *et al.* 2018). Snowmelt schemes, for example, include degree-day methods as used here or more complex energy balance approaches (Pohl *et al.* 2005; Corripio & López-Moreno 2017). Given the significance of snow accumulation and melt within the Lena, these different methods could simulate variable responses to the same climate change scenario.

Implications for the AMOC

Increases in mean discharge at Stolb dominate scenario results (10 of 12 GCM groups). Projections range between −5.3% (840 $m^3\ s^{-1}$) and +21.7% (3,440 $m^3\ s^{-1}$) with the group ensemble mean projecting an increase of 9.2% (1,458 $m^3\ s^{-1}$). These results can be used to provide estimates of potential changes in Eurasian runoff to the Arctic Ocean. The Lena, Yenisei and Ob contribute approximately 45% (~46,700 $m^3\ s^{-1}$) of the mean annual runoff to the Arctic (Ye *et al.* 2004). Changes in these inflows can be established if percentage changes for Stolb are applied to the downstream gauging station of each river (Igarka and Salekhard for Yenisei and Ob, respectively; R-Arctic.Net). These estimates assume regionally homogeneous climatic changes, the same hydrological responses to these changes within the Yenisei and Ob basins, and exclude future anthropogenic impacts. Whilst it is recognised that this is a simplification, the approach does enable an initial assessment of climate change driven modifications to Eurasian runoff to the Arctic for each of the 12 GCM groups. Increases of between 1,729 $m^3\ s^{-1}$ and 10,146 $m^3\ s^{-1}$ (1.7–10.1 mSv) (1 Sv = 1,000 mSv = $10^6\ m^3\ s^{-1}$) are projected for 10 groups (declines of 0.8–2.5 mSv for two) and the group ensemble mean (Figure 6).

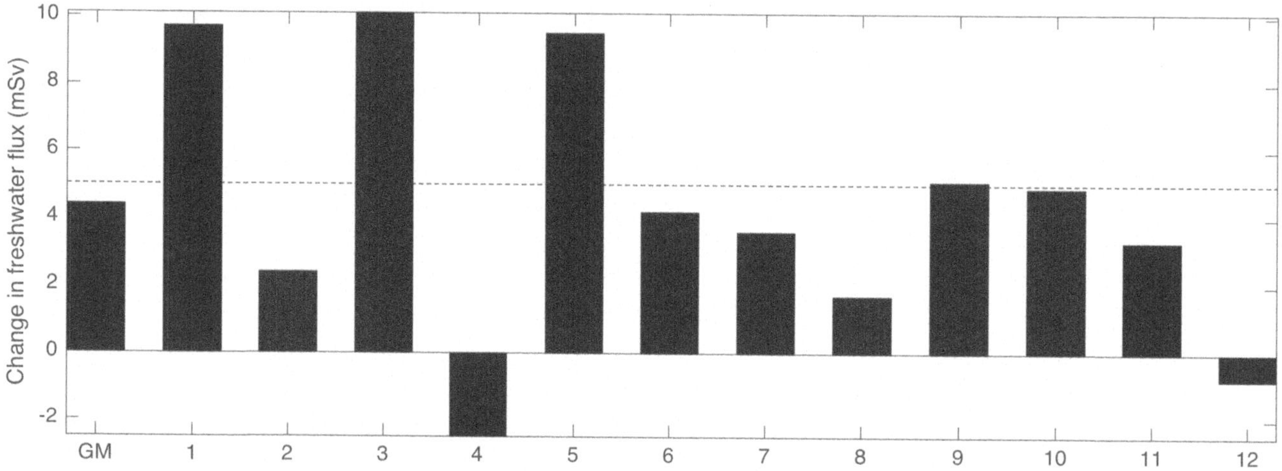

Figure 6 | Projected additional annual fluxes of freshwater from the three Eurasian Rivers (Lena, Yenisei and Ob) to the Arctic Ocean for each GCM group and the group ensemble mean (GM). The dashed line represents the minimum amount required to weaken AMOC (Schulz *et al.* 2007; Yang *et al.* 2016).

Peterson *et al.* (2002) suggested that an additional freshwater flux of between 60 mSv and 150 mSv would inhibit NADW formation. The changes summarised in Figure 6 are far below these values. However, sustained enhanced fluxes of 5–100 mSv could weaken convection (Schulz *et al.* 2007; Yang *et al.* 2016). Four of the GCM groups (1, 3, 5 and 9) project fluxes that cross the minimum threshold with projections for Group 10 coming close (Figure 6). Climate-induced changes in Eurasian river discharge under RCP4.5 may, therefore, produce freshwater fluxes capable of weakening AMOC. This extends the analysis of Shu *et al.* (2017), who found that enhanced runoff from all Arctic rivers under RCP8.5 could weaken AMOC, by suggesting that such a weakening may occur under a broader range of future climate conditions. Additionally, as Lena discharge shows increases with temperature (Gosling *et al.* 2017), it is likely that under the higher temperatures projected by RCP6.0 and RCP8.5, discharge may increase further, thus causing a more substantial weakening of AMOC. Therefore, whilst AMOC collapse may be improbable during the 21st century, it may be more likely in the future. Furthermore, the increases projected herein will be coupled with warmer temperatures (Thornalley *et al.* 2013), likely increased North American Arctic discharges (e.g. Arnell 2005; Shu *et al.* 2017), greater precipitation, and meltwater from the Greenland ice sheet (Vihma *et al.* 2016), which will all also act to reduce convection. These changes will increase the potential for a weakening of the AMOC that will have important implications for global climate.

CONCLUSION

A MIKE SHE/MIKE 11 model was used to investigate climate change impacts on discharge within the Lena River Basin for 12 genealogical-based GCM groups and the RCP4.5 scenario in 2071–2100. All groups projected basin-wide increases in precipitation, temperature and PET. However, the magnitudes of changes varied. Increases in mean annual discharge dominate with declines restricted to two groups. Seasonal shifts in the timing of snowmelt were simulated due to increases in temperature and precipitation during winter and spring. The application of projected changes to the three major Eurasian rivers suggests that AMOC weakening could potentially occur if enhanced freshwater inputs are sustained. When augmented by increases in other freshwater sources, and combined with higher temperature more groups may cross the threshold, increasing the likelihood of AMOC weakening by the end of the 21st century. Eurasian rivers alone could, therefore, play a significant role in altering this component of the Earth's climate system.

ACKNOWLEDGEMENTS

We thank two anonymous reviewers for their valuable comments and suggestions on earlier drafts of the manuscript.

REFERENCES

Allen, R., Pereira, L., Raes, D. & Smith, M. 1998 *Crop Evapotranspiration – Guidelines for Computing Crop Water Requirements*. Food and Agriculture Organisation, Rome.

Anandhi, A., Frei, A., Pierson, D. C., Schneiderman, E. M., Zion, M. D., Lounsbury, D. & Matonse, A. H. 2011 Examination of change factor methodologies for climate change impact assessment. *Water Resources Research* **47**, W03501, doi:10.1029/2010WR009104.

Andersen, J., Refsgaard, J. C. & Jensen, K. H. 2001 Distributed hydrological modelling of the Senegal River Basin – model construction and validation. *Journal of Hydrology* **247**, 200–214.

Arnell, N. W. 2005 Implications of climate change for freshwater inflows to the Arctic Ocean. *Geophysical Research Letters* **27**, 1183–1186.

Atwell, B., Kriedemann, P. & Turnbull, C. 1999 *Plants in Action: Adaptations in Nature, Performance in Cultivation*. MacMillan Education Australia, South Yarra.

Berezovskaya, S., Yang, D. & Hinzman, L. 2005 Long-term annual water balance analysis of the Lena River. *Global and Planetary Change* **48**, 84–95.

Broecker, W. S., Peteet, D. M. & Rind, D. 1985 Does the ocean-atmosphere system have more than one stable mode of operation? *Nature* **315**, 21–26.

Buckley, M. W. & Marshall, J. 2016 Observations, inferences, and mechanisms of the Atlantic Meridional Overturning Circulation: a review. *Reviews of Geophysics* **54**, 5–63.

Clark, P. U., Pisias, N. G., Stocker, T. F. & Weaver, A. J. 2002 The role of the thermohaline circulation in abrupt climate change. *Nature* **415**, 863–869.

Corripio, J. G. & López-Moreno, J. I. 2017 Analysis and predictability of the hydrological response of mountain catchments to heavy rain on snow events: a case study in the Spanish Pyrenees. *Hydrology* **4**, 20.

Dzhamalov, R., Krichevets, G. & Safronova, T. 2012 Current changes in water resources in Lena River basin. *Water Resources* **39**, 147–160.

Frieler, K. P., Lange, S. A., Piontek, F. D., Reyer, C. P. O. P., Schewe, J. K., Warszawski, L. N., Zhao, F. F., Chini, L., Denvil, S., Emanuel, K., Geiger, T., Halladay, K., Hurtt, G., Mengel, M., Murakami, D., Ostberg, S., Popp, A., Riva, R., Stevanovic, M., Suzuki, T., Volkholz, J., Burke, E., Ciais, P., Ebi, K., Eddy, T. D., Elliott, J., Galbraith, E., Gosling, S. N., Hattermann, F., Hickler, T., Hinkel, J., Hof, C., Huber, V., Jägermeyr, J., Krysanova, V., Marcé, R., Müller Schmied, H., Mouratiadou, I., Pierson, D., Tittensor, D. P., Vautard, R., van Vliet, M., Biber, M. F., Betts, R. A., Bodirsky, B. L., Deryng, D., Frolking, S., Jones, C. D., Lotze, H. K., Lotze-Campen, H., Sahajpal, R., Thonicke, K., Tian, H. & Yamagata, Y. 2017 Assessing the impacts of 1.5°C global warming – simulation protocol of the inter-sectoral impact model intercomparison project (ISIMIP2b). *Geoscientific Model Development* **10**, 4321–4345.

Gautier, E., Dépret, T., Costard, F., Virmoux, C., Deforov, A., Grancher, D., Konstantinov, P. & Brunstein, D. 2018 Going with the flow: hydrologic response of the middle Lena River (Siberia) to the climate variability and change. *Journal of Hydrology* **557**, 475–488.

Gelfan, A., Gustafsson, D., Motovilov, Y., Arheimer, B., Kalugin, A., Krylenko, I. & Lavrenov, A. 2017 Climate change impact on the water regime of two great Arctic rivers: modelling and uncertainty issues. *Climatic Change* **141**, 499–515.

Gosling, S. N. 2013 The likelihood and potential impact of future change in the large-scale climate-earth system on ecosystem services. *Environmental Science and Policy* **27**, 15–31.

Gosling, S. N., Taylor, R. G., Arnell, N. W. & Todd, M. C. 2011 A comparative analysis of projected impacts of climate change on river runoff from global and catchment-scale hydrological models. *Hydrology and Earth System Sciences* **15**, 279–294.

Gosling, S. N., Zaherpour, J., Mount, N. J., Hattermann, F. F., Dankers, R., Arheimer, B., Breuer, L., Ding, J., Haddeland, I., Kumar, R., Kundu, D., Liu, J., Van Griensven, A., Veldkamp, T. I. E., Vetter, T., Wang, X. & Zhang, X. 2017 A comparison of changes in river runoff from multiple global and catchment-scale hydrological models under global warming scenarios of 1 °C, 2 °C and 3 °C. *Climatic Change* **141**, 577–595.

Groisman, P., Koknaeva, V., Belokrylova, T. & Karl, T. 1991 Overcoming biases of precipitation measurement: a history of the USSR experience. *Bulletin of the American Meteorological Society* **72**, 1725–1733.

Hargreaves, G. H. & Samani, Z. A. 1985 Reference crop evapotranspiration from temperature. *Applied Engineering in Agriculture* **1**, 96–99.

Harris, I., Jones, P., Osborn, T. & Lister, D. 2014 Updated high-resolution grids of monthly climatic observations – the CRU TS3.10 dataset. *International Journal of Climatology* **34**, 623–642.

Hattermann, F. F., Krysanova, V., Gosling, S. N., Dankers, R., Daggupati, P., Donnelly, C., Flörke, M., Huang, S., Motovilov, Y., Buda, S., Yang, T., Müller, C., Leng, G., Tang, Q., Portmann, F. T., Hagemann, S., Gerten, D., Wada, Y., Masaki, Y., Alemayehu, T., Satoh, Y. & Samaneigo, L. 2017 Cross-scale intercomparison of climate change impacts simulated by regional and global hydrological models in eleven large river basins. *Climatic Change* **141**, 561–576.

Hattermann, F. F., Vetter, T., Breuer, L., Su, B., Daggupati, P., Donnelly, C., Fekete, B., Flörke, F., Gosling, S. N., Hoffmann, P., Liersch, S., Masaki, Y., Motovolov, Y., Müller, C., Samaniego, L., Stacke, T., Wada, Y., Yang, T. & Krysnaova, V. 2018 Sources of uncertainty in hydrological climate impact assessment: a cross-scale study. *Environmental Research Letters* **13**, 015006.

Hawkins, E., Smith, R. S., Allison, L. C., Gregory, J. M., Woollings, T. J., Pohlmann, H. & De Cuevas, B. 2011 Bistability of the Atlantic overturning circulation in a global climate model and links to ocean freshwater transport. *Geophysical Research Letters* **38**, L10605.

Henriksen, H. J., Troldborg, L., Nyegaard, P., Sonnenborg, T. O., Refsgaard, J. C. & Madsen, B. 2003 Methodology for construction, calibration and validation of a national hydrological model for Denmark. *Journal of Hydrology* **280**, 52–71.

Ho, J. T., Thompson, J. R. & Brierley, C. 2016 Projections of hydrology in the Tocantins-Araguaia Basin, Brazil: uncertainty assessment using the CMIP5 ensemble. *Hydrological Sciences Journal* **61**, 551–567.

Holmes, R. M., McClelland, J. W., Peterson, B. J., Tank, S. E., Bulygina, E., Eglinton, T. I., Gordeev, V. V., Gurtovaya, T. Y., Raymond, P. A., Repeta, D. J., Staples, R., Striegl, R. G., Zhulidov, A. V. & Zimov, S. A. 2012 Seasonal and annual fluxes of nutrients and organic matter from large rivers to the Arctic Ocean and surrounding seas. *Estuaries and Coasts* **35**, 369–382.

Immerzeel, W., Pellicciotti, F. & Shrestha, A. 2012 Glaciers as a proxy to quantify the spatial distribution of precipitation in the Hunza Basin. *Mountain Research and Development* **32**, 30–38.

Ji, X. & Luo, Y. 2013 The influence of precipitation and temperature input schemes on hydrological simulations of a snow and glacier melt dominated basin in Northwest China. *Hydrology and Earth System Sciences* **10**, 807–853.

Keller, K., Tan, K., Morel, F. M. M. & Bradford, D. F. 2000 Preserving the ocean circulation: implications for climate policy. *Climatic Change* **47**, 17–43.

Kingston, D. G., Thompson, J. R. & Kite, G. 2011 Uncertainty in climate change projections of discharge for the Mekong River Basin. *Hydrology and Earth System Sciences* **15**, 1459–1471.

Kirtman, B., Power, S. B., Adedoyin, J. A., Boer, G. J., Bojariu, R., Camilloni, I., Doblas-Reyes, F. J., Fiore, A. M., Kimoto, M., Meehl, G. A., Prather, M., Sarr, A., Schär, C., Sutton, R., van Oldenborgh, G. J., Vecchi, G. & Wang, H. J. 2013 Near-term climate change: projections and predictability. In: *Climate Change 2013: The Physical Science Basis. Contribution of Working Group I to the Fifth Assessment Report of the Intergovernmental Panel on Climate Change* (T. F. Stocker, D. Qin, G.-K. Plattner, M. Tignor, S. K. Allen, J. Boschung, A. Nauels, Y. Xia, V. Bex & P. M. Midgley, eds). Cambridge University Press, Cambridge, pp. 953–1028.

Köhler, P., Joos, F., Gerber, S. & Knutti, R. 2005 Simulated changes in vegetation distribution, land carbon storage, and atmospheric CO_2 in response to a collapse of the North Atlantic thermohaline circulation. *Climate Dynamics* **25**, 689–708.

Krysanova, V., Vetter, T., Eisner, S., Huang, S., Pechlivanidis, I., Strauch, M., Gelfan, A., Kumar, R., Aich, V., Arheimer, B. & Chamorro, A. 2017 Intercomparison of regional-scale hydrological models and climate change impacts projected for 12 large river basins worldwide – a synthesis. *Environmental Research Letters* **12**, 105002.

Krysanova, V., Donnelly, C., Gelfan, A., Gerten, D., Arheimer, B., Hattermann, F. & Kundzewicz, Z. W. 2018 How the performance of hydrological models relates to credibility of projections under climate change. *Hydrological Sciences Journal* **63**, 696–720.

Kuhlbrodt, T., Rahmstorf, S., Zickfeld, K., Vikebø, F., Sundby, S., Hofmann, M., Link, P., Bondeau, A., Cramer, W. & Jaeger, C. 2009 An integrated assessment of changes in the thermohaline circulation. *Climatic Change* **96**, 489–537.

Liu, B. & Yang, D. 2011 Siberian Lena River heat flow regime and change. *Cold Region Hydrology in a Changing Climate* **346**, 71–76.

Manabe, S. & Stouffer, R. J. 1988 Two stable equilibria of a coupled ocean-atmosphere model. *Journal of Climate* **1**, 871–884.

Maxino, C. C., Mcavaney, B. J., Pitman, A. J. & Perkins, S. E. 2008 Ranking the AR4 climate models over the Murray-Darling Basin using simulated maximum temperature, minimum temperature and precipitation. *International Journal of Climatology* **28**, 1097–1112.

Overeem, I. & Syvitski, J. P. M. 2010 Shifting discharge peaks in Arctic rivers, 1977–2007. *Geografiska Annaler: Series A, Physical Geography* **92**, 285–296.

Overland, J. E., Wang, M., Walsh, J. E. & Stroeve, J. C. 2013 Future Arctic climate changes: adaptation and mitigation time scales. *Earth's Future* **2**, 68–74.

Peterson, B. J., Holmes, R. M., McClelland, J. W., Vörösmarty, C. J., Lammers, R. B., Shiklomanov, A. I., Shiklomanov, I. A. & Rahmstorf, S. 2002 Increasing river discharge to the Arctic Ocean. *Science* **298**, 2171–2173.

Pirtle, Z., Meyer, R. & Hamilton, A. 2010 What does it mean when climate models agree? A case for assessing independence among general circulation models. *Environmental Science and Policy* **13**, 351–361.

Pohl, S., Davison, B., Marsh, P. & Pietroniro, A. 2005 Modelling spatially distributed snowmelt and meltwater runoff in a small Arctic catchment with a hydrology land-surface scheme (WATCLASS). *Atmosphere-Ocean* **43**, 193–211.

Rawlins, M. A., Willmott, C. J., Shiklomanov, A., Linder, E., Frolking, S., Lammers, R. B. & Vörösmarty, C. J. 2006 Evaluation of trends in derived snowfall and rainfall across Eurasia and linkages with discharge to the Arctic Ocean. *Geophysical Research Letters* **33**, L07403.

Refsgaard, J. C., Storm, B. & Clausen, T. 2010 Système Hydrologique Europeén (SHE): review and perspectives after 30 years development in distributed physically-based hydrological modelling. *Hydrology Research* **41**, 355–377.

Robinson, A. J. 2018 *Uncertainty in Hydrological Scenario Modelling: An Investigation Using the Mekong River Basin, SE Asia.* PhD Thesis, UCL, London.

Schulz, M., Prange, M. & Klocker, A. 2007 Low-frequency oscillations of the Atlantic Ocean meridional overturning circulation in a coupled climate model. *Climate of the Past* **3**, 97–107.

Shiklomanov, I. A., Shiklomanov, A. I., Lammers, R. B., Peterson, B. J. & Vörösmarty, C. J. 2000 The dynamics of river water inflow to the Arctic Ocean. In: *The Freshwater Budget of the Arctic Ocean: Proceedings of the NATO Advanced Research Workshop* (E. L. Lewis, E. P. Jones & P. Lemke, eds). Kluwer, Dordrecht, pp. 281–296.

Shu, Q., Qiao, F., Song, Z. & Xiao, B. 2017 Effect of increasing Arctic river runoff on the Atlantic meridional overturning circulation: a model study. *Acta Oceanologica Sinica* **36**, 59–65.

Thompson, J. R., Refstrup-Sørenson, H., Gavin, H. & Refsgaard, A. 2004 Application of the coupled MIKE SHE/MIKE 11 modelling system to a lowland wet grassland in Southeast England. *Journal of Hydrology* **293**, 151–179.

Thompson, J. R., Green, A. J., Kingston, D. J. & Gosling, S. N. 2013 Assessment of uncertainty in river flow projections for the Mekong River using multiple GCMs and hydrological models. *Journal of Hydrology* **486**, 1–30.

Thompson, J. R., Green, A. J. & Kingston, D. J. 2014 Potential evapotranspiration-related uncertainty in climate change impacts on river flow: an assessment for the Mekong River. *Journal of Hydrology* **510**, 259–279.

Thompson, J. R., Crawley, A. & Kingston, D. G. 2017a Future river flows and flood extent in the Upper Niger and Inner Niger Delta: GCM-related uncertainty using the CMIP5 ensemble. *Hydrological Sciences Journal* **62**, 2239–2265.

Thompson, J. R., Iravani, H., Clilverd, H. M., Sayer, C. D., Heppell, C. M. & Axmacher, J. C. 2017b Simulation of the hydrological impacts of climate change on a restored floodplain. *Hydrological Sciences Journal* **62**, 2482–2510.

Thornalley, D. J. R., Blaschek, M., Davies, F. J., Praetorius, S., Oppo, D. W., McManus, J. F., Hall, I. R., Kleiven, H., Renssen, H. & McCave, I. N. 2013 Long-term variations in Iceland-Scotland overflow strength in the Holocene. *Climate of the Past* **9**, 2073–2084.

UNFCCC 2015 Adoption of the Paris Agreement. In *21st Conference of the Parties*. United Nations, Paris.

Veldkamp, T. I. E., Zhao, F., Ward, P. J., de Moel, H., Aerts, J. C. J. H., Muller Schmied, H., Portmann, F. T., Masaki, Y., Pokhrel, Y., Liu, X., Satoh, Y., Gerten, D., Gosling, S. N., Zaherpour, J. & Wada, Y. 2018 Human impact parameterisations in global hydrological models improve estimates of monthly discharges and hydrological extremes: a multi-model validation study. *Environmental Research Letters* **13**, 055008.

Vellinga, M. & Wood, R. A. 2008 Impacts of thermohaline circulation shutdown in the twenty-first century. *Climatic Change* **91**, 43–63.

Vihma, T., Screen, J., Tjerström, M., Newton, B., Zhang, X., Popova, V., Deser, C., Holland, M. & Prowse, T. 2016 The atmospheric role in the Arctic water cycle: a review on processes, past and future changes, and their impacts. *Journal of Geophysical Research* **121**, 586–620.

Walvoord, M. A. & Kurylyk, B. L. 2016 Hydrologic impacts of thawing permafrost – a review. *Vadose Zone Journal* **15**, 1–20.

Walvoord, M. A. & Striegl, R. G. 2007 Increased groundwater to stream discharge from permafrost thawing in the Yukon River basin: potential impacts on lateral export of carbon and nitrogen. *Geophysical Research Letters* **34**, 1–6.

Wang, Q., Fan, X. & Wang, M. 2016 Evidence of high-elevation amplification versus Arctic amplification. *Scientific Reports* **6**, 19219.

Woo, M.-K., Thorne, R., Szeto, K. & Yang, D. 2008 Streamflow hydrology in the boreal region under the influences of climate and human interference. *Philosophical Transactions: Biological Sciences* **363**, 2251–2260.

Yang, Q., Dixon, T., Myers, P., Bonin, J., Chambers, D. & Van Den Broeke, M. 2016 Recent increases in Arctic freshwater flux affects Labrador Sea convection and Atlantic overturning circulation. *Nature Communications* **7**, 1–7.

Ye, B., Yang, D. & Kane, D. L. 2003 Changes in Lena River streamflow hydrology: human impacts versus natural variations. *Water Resources Research* **39**, 1200.

Ye, H., Yang, D., Zhang, T., Shang, X., Ladochy, S. & Ellison, M. 2004 The impact of climatic conditions on seasonal river discharges in Siberia. *Journal of Hydrometeorology* **5**, 286–295.

Zaherpour, J., Mount, N., Gosling, S. N., Dankers, R., Eisner, S., Gerten, D., Liu, X., Masaki, Y., Schmied, H. M., Tong, Q. & Wada, Y. 2019 Exploring the value of machine learning for weighted multi-model combination of an ensemble of global hydrological models. *Environmental Modelling and Software* **114**, 112–128.

First received 29 November 2018; accepted in revised form 9 July 2019. Available online 2 August 2019

Hydrological functioning of cattle ranching impoundments in the Dry Chaco rangelands of Argentina

Patricio N. Magliano, David Mindham, Wlodek Tych, Francisco Murray, Marcelo D. Nosetto, Esteban G. Jobbágy, Marcos J. Niborski, Mariana C. Rufino and Nick A. Chappell

ABSTRACT

Rainwater harvesting and associated storage is essential for cattle ranching in the drylands of Argentina and elsewhere. This is the first study to attempt to quantify the hydrological inflows and losses from rainwater harvesting impoundments. To address the direct effect of cattle within impoundments, a typical cattle-affected impoundment was instrumented and compared with that of a similar impoundment but without cattle access. Analysis of the storage dynamics with reference to the controlling variables demonstrated the highly episodic nature of the generation of infiltration-excess overland flow that recharged the impoundments. The impoundments experienced 43 and 35% of storage loss to open-water-evaporation for the cattle-affected and control impoundments, respectively. Critically, the cattle-effected impoundment lost only 15% of storage to leakage (after cattle consumption was taken into account), while the control lost 65% of its water to basal leakage. Indeed systems modelling of the rainfall-storage dynamics showed that the cattle-affected impoundment, despite consumption by 300 cows, maintained water in the impoundment (per a unit input of rainfall) for longer than the control (a 65- versus 25-day residence time). These results highlight the unintended beneficial effect of cattle trampling on the floor of the impoundment reducing leakage losses.

Key words | arid, ecohydrology, grazing, livestock, rainwater harvesting, runoff

Patricio N. Magliano (corresponding author)
Francisco Murray
Marcelo D. Nosetto
Esteban G. Jobbágy
Grupo de Estudios Ambientales – IMASL,
Universidad Nacional de San Luis & CONICET,
Ejército de los Andes 950, D5700HHW San Luis,
Argentina
E-mail: *pnmagliano@gmail.com*

Patricio N. Magliano
Departamento de Bioquímica y Ciencias Biológicas, Facultad de Química, Bioquímica y Farmacia,
Universidad Nacional de San Luis (UNSL),
Ejército de los Andes 950, D5700HHW San Luis,
Argentina

Patricio N. Magliano
David Mindham
Wlodek Tych
Mariana C. Rufino
Nick A. Chappell
Lancaster Environment Centre,
Lancaster University,
Lancaster, UK

Francisco Murray
Instituto Nacional de Tecnología Agropecuaria (INTA),
AER San Luis, Ruta 20 km 4.5, D5700HHW San Luis,
Argentina

Marcelo D. Nosetto
Cátedra de Climatología,
Facultad de Ciencias Agropecuarias (UNER),
Entre Ríos, Argentina

Marcos J. Niborski
Cátedra de Manejo y Conservación de Suelos, Facultad de Agronomía,
Universidad de Buenos Aires (UBA),
Avda. San Martín 4453, C1417DSE Ciudad Autónoma de Buenos Aires, Argentina

INTRODUCTION

Extensive livestock production occupies one-third of the global land area and represents a key economic activity in low- and middle-income countries (Schimel 2010; Robinson *et al.* 2014). Given that most livestock systems are located in arid and semi-arid regions (hereafter drylands), water is the main constraint to production due to the shortage of palatable forage and limited access to drinking water for the cattle (Evenari *et al.* 1971; UNEP 2009). In the case of

doi: 10.2166/nh.2019.149

forage shortage, there are several management practices to consider, such as adjusting stocking rates (Ganskopp & Bohnert 2009; Fernández *et al.* 2015). However, water supply represents a crucial limitation for cattle survival (Ganskopp 2001; Bailey 2005) and it has been historically overcome by capturing infiltration-excess overland flow during rain events, so called 'rainwater harvesting' (Lavee *et al.* 1997; Oweis & Hachum 2009; Denison & Wotshela 2012). Huxman *et al.* (2004), Newman *et al.* (2006) and Jobbágy *et al.* (2008) have shown that this is feasible in dry and flat sedimentary landscapes where infiltration-excess overland flow may represent less than 5% of rainfall inputs. Infiltration-excess overland flow is surface flow on slopes (rather than channels) created when short-term rainfall intensities locally (in time and space) exceed the infiltration capacity of the soil (equivalent to the saturated hydraulic conductivity of the soil surface).

By definition, rainwater harvesting comprises of capturing, storing and then using infiltration-excess overland flows for water supply (Critchley *et al.* 1991; Fox & Rockström 2000). Most rainwater harvesting systems have a well-defined area upslope of an impoundment, where infiltration-excess overland flow is generated (Boers & Ben-Asher 1982; Pandey *et al.* 2003). Each impoundment may be a natural topographic depression or man-made excavation with embankments (Glendenning *et al.* 2012; Adham *et al.* 2016). In very flat landscapes it can be difficult to determine the exact contributory area of an impoundment because it is hard to define topographic divide and because of the enhanced role of preferential pathways of vehicle roads or cattle trails that may connect distal areas with the impoundment (Pandey *et al.* 2003; Magliano *et al.* 2015c). The generation of infiltration-excess overland flow is, however, enhanced by the presence of the high-intensity convective rainfall that characterizes semi-arid regions.

The Dry Chaco rangelands of northern Argentina (also known as Arid Chaco) is ~10 million hectares of sedimentary plain dominated by native dry woodlands (Cabrera 1976; Oyarzabal *et al.* 2018). In this region, rainwater harvesting systems used for cattle ranching are typically composed of a man-made impoundment associated with a contributory area crossed by cattle trails and vehicle roads (Karlin *et al.* 2013; Magliano *et al.* 2015c). Impoundments, with typical dimensions of approximately 50×80 m, consist of arched embankments built 2–4 m above the ground with an open section that allows entry of the infiltration-excess overland flow. Cattle drink water directly from within the impoundment so trampling the floor of the impoundment (in addition to trampling of the access points and contributory catchment) may be important. Cattle trails include networks of tracks characterised by bare and compacted ground that converges onto the open section of the impoundment. They are created by cattle as they move daily from grazing sites to the impoundments (George *et al.* 2004; Ganskopp & Bohnert 2009; Richardson & Rejmánek 2011; Karlin *et al.* 2013). These trails are often associated with an overgrazed surface zone close to the open side of the impoundment that is sometimes called the 'piosphere' or zone of focused impact of grazing animals on soils and vegetation (Macchi & Grau 2012; Chillo *et al.* 2015). Vehicle tracks are also present in the contributory catchments and are typically unpaved, slightly depressed roadways. There is some evidence that undisturbed Dry Chaco woodlands do not generate infiltration-excess overland flow that is able to travel more than a few metres (Aguilera *et al.* 2003; Kunst *et al.* 2003; Magliano *et al.* 2016; Magliano *et al.* 2015a), highlighting the importance of the cattle trails and vehicle tracks to generate the infiltration-excess overland flow that fills the impoundments built for cattle water supply.

Despite its great importance to cattle production in Dry Chaco rangelands, there is little scientific knowledge about the hydrological functioning of these impoundments and the biophysical controls on their water fluxes. Typically, there is one rainwater harvesting system every 1,000 ha within the cattle-grazed rangeland, that each collect only a tiny fraction of the rain falling in their contributory areas (Magliano *et al.* 2015c). Although rainwater harvesting systems make cattle production possible in the region, they are not considered efficient systems because of their perceived large water losses. During dry years, or at the end of the dry season, impoundments often run out of water generating serious problems for cattle ranchers. While local ranchers believe that water depletion in impoundments results from a combination of consumption by cattle and evaporation, the preliminary study of Magliano *et al.* (2015c) suggested that leakage through the impoundment floor may be an important water loss. Detailed understanding of the hydrological functioning of example impoundments is, therefore, a potential first step to improving the design and management of these rainwater harvesting systems in the Dry Chaco of Argentina and similar systems in other rangelands worldwide.

This study explored the hydrological functioning of two instrumented impoundments built to hold harvested rainwater in the Dry Chaco rangelands of Argentina (Figure 1). Two particular impoundments were selected for this first pioneering study because of the potential importance of the combined effects of cattle consumption of the impoundment water and the effects of trampling of the impoundment floor during cattle access for drinking. One impoundment ('cattle impoundment') was selected, as cattle accessed that impoundment to drink (Figure 1). The second impoundment did not have cattle access or water consumption (though it did have compacted tracks within its contributory area) and was selected as a reference or 'control' to permit the study of the effects of cattle presence/absence within an impoundment.

The first objective of the study was to quantify the storage dynamics of the two instrumented impoundments over the water year. The second objective was to estimate, for the first time: (1) the temporal dynamics of the additions to storage from infiltration-excess overland flow from the local rainfall, and (2) a decomposition of the losses from storage attributed separately to open water evaporation, cattle drinking and leakage through the impoundment floor. As the most critical attribute of the cattle-drinking impoundments is maintenance of storage for the longest period possible (i.e. security from water supply failure), the dynamic relationship between the unique synchronous rainfall and impoundment storage data were analysed using a systems modelling approach. This final objective being to quantify the residence time of a unit input of rainfall within the two impoundments (i.e. how long storage is maintained for standardised rainfall inputs), one with cattle consumption and trampling, one without.

METHODS

Study site

Extensive cattle ranching represents the most important economic activity of the Dry Chaco rangelands of Argentina. Drinking water for cattle is supplied by rainwater harvesting systems, while native grasses and woody species provide the forage (Aguilera 2003). This study was undertaken within a 20,000 ha farm located in the centre of the San Luis province of Argentina (33.5°S, 66.5°W; Figure 1) and was managed by Ser Beef S.A. Typical stocking rates

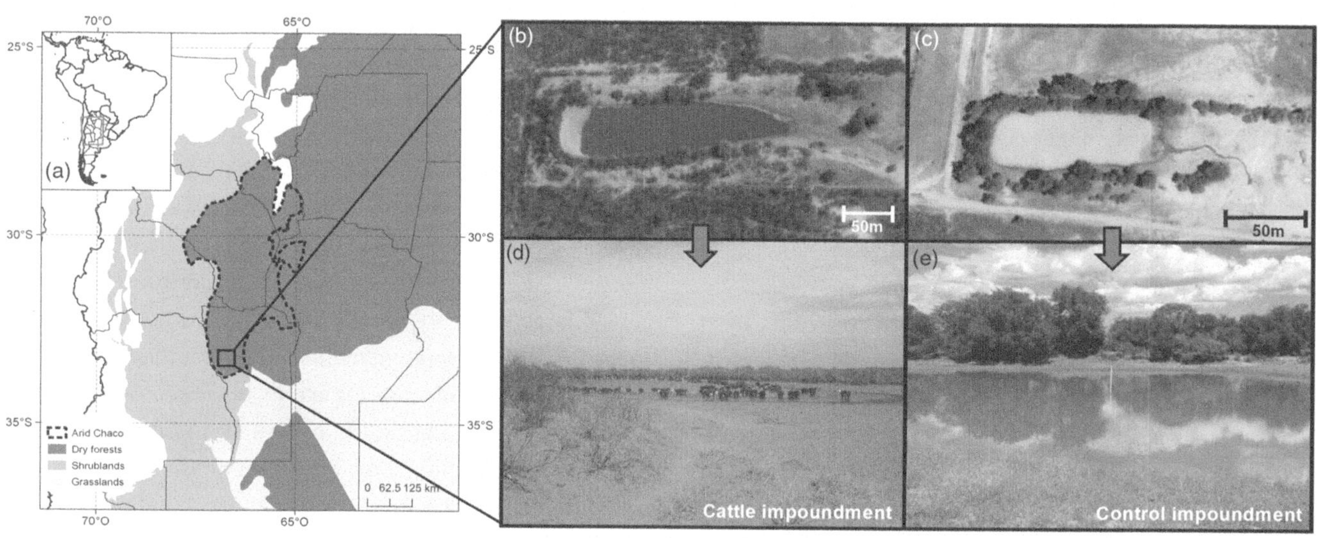

Figure 1 | Study site in central Argentina. (a) The map on the left shows the Dry Chaco rangelands (Arid Chaco, ~10 million of hectares), while (b) and (d) are images of the instrumented impoundment used for cattle water supply (cattle impoundment), and (c) and (e) are the control impoundment. Images (b) and (c) are publically available 'Quickbird' images from Google Earth, while the lead author took photographs (d) and (e).

were 0.15 calving units per ha (Ser Beef S.A. personal communication).

The vegetation is dominated by native dry woodlands with isolated stands of altered vegetation that include rainfed pastures of *Cenchrus ciliaris* and *Eragrostis curvula* and some localised irrigated plots of maize, sorghum and soybean (Marchesini *et al.* 2015; Steinaker *et al.* 2016). The native dry woodlands are dominated by two tree species: *Prosopis flexuosa* and *Aspidosperma quebracho-blanco*. The understorey includes several shrub species such as *Larrea divaricata* and *Senna aphylla*, and perennial grasses such as *Stipa eriostachya* and *Aristida mendocina* (García *et al.* 2017). The mean leaf area index of the dry woodland approaches 1 and mean incident radiation at the ground level is 47.5% (García *et al.* 2017).

The soils are well drained and derived from fine loessic sediments deposited throughout the Holocene with some alluvial reworking (Iriondo 1993; Tripaldi *et al.* 2013). Soils are Typic Torriorthents with 53% sand, 15% clay, and 1.4% organic matter in the top 10 cm of the profile (Peña Zubiate *et al.* 1998; Pennington *et al.* 2000). Topography is gentle with slopes of <1.5%. Saturated hydraulic conductivity values around 140 mm h^{-1} have been measured for the surface soil in nearby undisturbed, native dry woodlands, with dry bulk density and 'water holding capacity' values of approximately 1.2 g cm^{-3} and 20.6%, respectively (Magliano *et al.* 2017b). The regional water table is approximately 30 m deep at the study site (Ser Beef S.A. unpublished data).

The climate is semiarid with a long-term mean annual rainfall of only 429 ± 83 mm (mean and standard deviation), distributed in 40 ± 12 events per year (Magliano *et al.* 2015b). Rainfall has a strong seasonal pattern. Some 85% of annual rainfall occurs in the wet season of September–March. During the wet season, there are more rainfall events of larger size and so more likely to generate infiltration-excess overland flow. The mean annual potential evapotranspiration is approximately 1,500 mm $year^{-1}$ (New *et al.* 2002) and reaches maximum daily values of approximately 8.5 mm day^{-1} in the summer (Magliano *et al.* 2017a). The mean annual temperature is 17.8 °C. The hottest and coldest months have mean temperatures of 24.8 (January) and 10.3 °C (July), respectively. Some 38 ground frost events are observed each year, mostly in the cold dry season (New *et al.* 2002).

Field monitoring of control and cattle-affected impoundments

Imagery of the 'cattle impoundment' is given in Figure 1(b) and 1(d), and for the 'control impoundment' (with no cattle ingress) in Figure 1(c) and 1(e). At each impoundment, the water level was monitored using a data-logged pressure transducer (depth resolution ±2 mm; Hobo, Onset Corporation, Bourne, MA, USA). As the impoundments have a bowl shape, sensors were installed at the base of the deepest point, and the water level was recorded every 30 min. All monitoring was undertaken between 10th November 2010 and 4th January 2012, totalling 421 days of measurements. A detailed bathymetric survey of both impoundments was undertaken so that the daily-average water levels could be converted into values of daily water volume (Equations (1) and (2)) and daily water surface area (Equations (3) and (4)). Further details of the water volume calculations using an adapted fitting curve method applied to the high-resolution digital elevation model are presented in Magliano *et al.* (2015c). The derived level to volume relationships for the cattle and control impoundments are given in Equations (1) and (2), respectively:

$$V = 1335.5\,h^2 - 125.8\,h \tag{1}$$

$$V = 1033.9\,h^2 - 181.5\,h \tag{2}$$

where V is the water volume (m^3) and h is the water level (m), and level to water surface area relations are given in Equations (3) and (4), respectively:

$$S = 2586.7\,h \tag{3}$$

$$S = 1993.3\,h \tag{4}$$

where S is water surface (m^2) and h is water level (m)

Rainfall was measured with a tipping-bucket gauge installed 1 km from the impoundments (resolution = 0.1 mm per tip; Vantage Pro2 weather station, Davis Instruments, USA), and accumulated tips recorded every 30 min. The annual average rainfall recorded by the tipping-bucket gauge was checked against those data from a nearby storage rain gauge. Meteorological records of solar radiation, temperature, humidity and wind speed from the same Vantage Pro2 weather station were used to estimate open water evaporation using an adapted Penman method (following Allen *et al.* 1998).

Derivation of infiltration-excess inputs and evaporation, consumption and leakage outputs of the impoundments from the storage dynamics and associated data

Water loss from the two impoundments by the combined effect of open water evaporation, drainage through the impoundment floor and consumption by cattle, was derived on a daily time-step:

$$Y = (x_n - x_{n-1}), \text{ where } (x_n - x_{n-1}) < 0 \tag{5}$$

where a negative Y is daily water loss (m^3 day^{-1}), x is water volume at n day. In the case of $Y > 0$, this equates to very short episodes of recharge of the impoundment (m^3) by infiltration-excess overland flow from the surrounding 'rainwater harvesting' catchment.

Daily water loss was decomposed into estimates of the three component fluxes (evaporation, cattle consumption and drainage through the impoundment floor), using Equations (6)–(8), respectively:

$$y_e = x_1^* S, \text{ where } Y < 0 \tag{6}$$

where y_e is daily estimate of the volume of water lost by evaporation from the impoundment (m^3), x_1 is daily evaporation depth estimate by the adapted Penman method of Allen *et al.* (1998; m) and S is water surface area of the impoundment (m^2):

$$y_c = x_2^* n, \text{ where } Y < 0 \tag{7}$$

where y_c is daily cattle water consumption volume from the 'cattle impoundment' (m^3), x_2 is the daily rate of consumption estimated to be 0.05 m^3 per cow for dryland sites (Schlink *et al.* 2010) and n is the number of cattle accessing the 'cattle impoundment' each day (300 cows: Ser Beef S.A. personal communication). The cattle access the studied impoundment eight months per year; in this study, the access started on 15th February 2011 (Ser Beef S.A. personal communication). Consequently, the estimate of the drainage through the floor of the impoundment is the total water loss minus the evaporation and consumption:

$$y_i = Y - y_e - y_c, \text{ where } Y < 0 \tag{8}$$

where y_i is daily volume of infiltration into the floor of the wetted area of the impoundment (m^3), Y is daily volume losses from the rate of change of water level (m^3), y_e is daily evaporation volume (m^3) and y_c is daily cattle consumed water volume (m^3).

The overall input–output water balance of each impoundment over the 421-day period, taking into account the infiltration-excess overland flow inputs of the rainwater harvesting system (m^3) and the losses (evaporation, cattle consumption and drainage through the impoundment floor, m^3), was then computed with Equation (9):

$$y = \sum a - \sum b \tag{9}$$

where y is the daily volume balance, a is an impoundment input and b is an impoundment loss. To explore the dynamics these daily values were expressed as a percentage of the maximum observed storage (6,140 and 4,102 m^3 for cattle and control impoundment, respectively).

Systems analysis of the input-storage dynamics of the impoundments

Rainwater harvesting is the dominant strategy for supplying drinking water to cattle in the Dry Chaco rangelands of Argentina. The main concern of ranchers is to have enough water at the end of the dry season, i.e. August (Figure 2(a)). Consequently, a key hydrological characteristic of the impoundments constructed for cattle water supply is how long a unit input of rainfall can be maintained in an impoundment.

To quantify how long storage can be maintained following a unit input of rainfall, a transfer-function approach has been used to identify the relationship between rainfall intensity and impoundment storage. The approach used was the

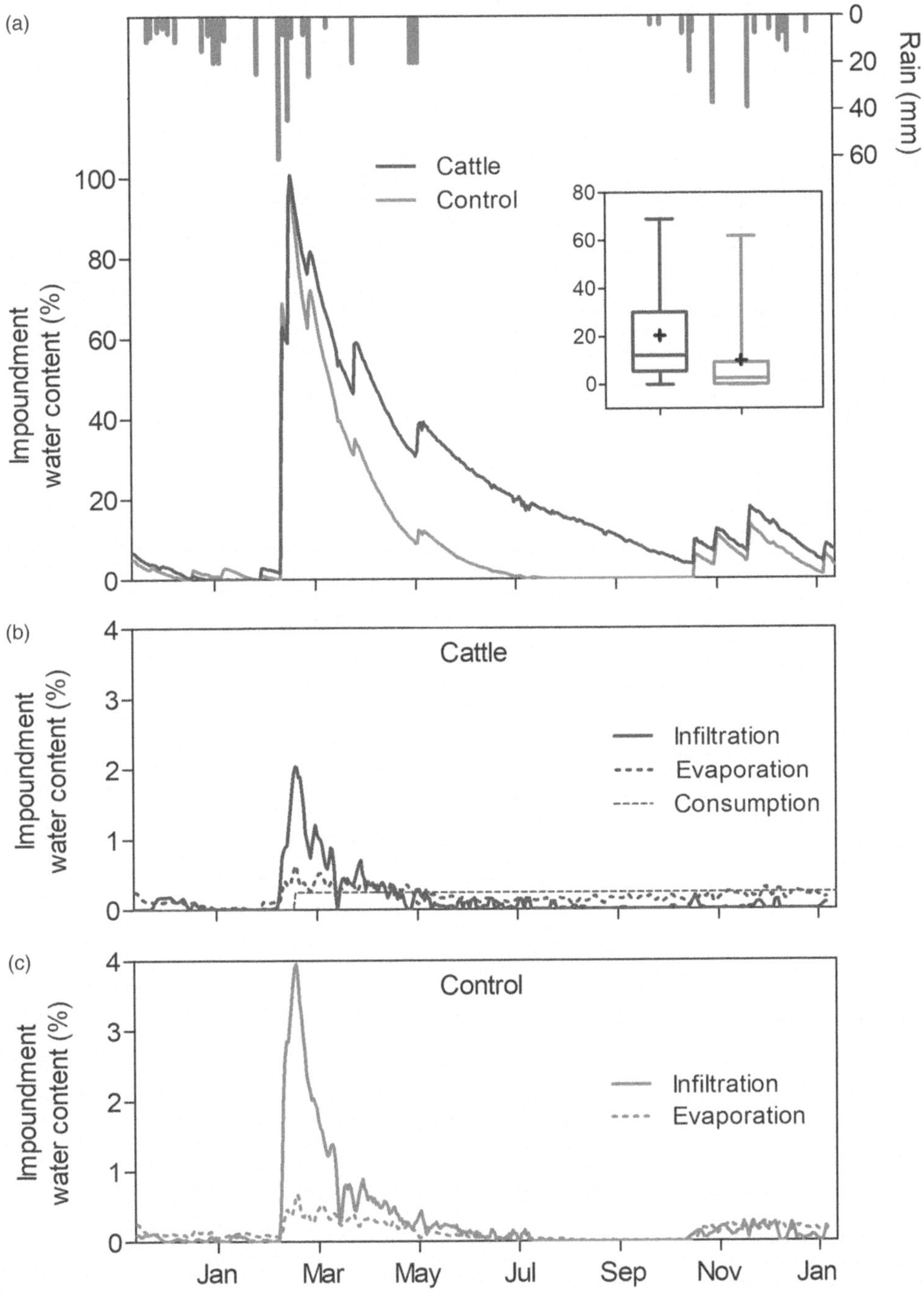

Figure 2 | Temporal dynamics of water content and derived losses for cattle and control impoundments, where: (a) water content dynamics, with the inset showing the distribution of the data; (b) dynamics of the derived estimates of evaporation, infiltration/leakage and consumption in the cattle impoundment; and (c) dynamics of the same variables for control impoundment.

Refined Instrumental Variable in Continuous-time (RIVC) algorithm of Young (2015) which is one of the algorithms that form the CAPTAIN Toolbox for Matlab™ (Taylor *et al.* 2007).

In summary, the RIVC algorithm implements an iterative instrumental variable method for estimation of general *Transfer Functions* that capture the dynamic relationship between an input (here rainfall, R) and an output variable (here impoundment water volume, V) that can be written as:

$$V = \frac{\beta}{s-\alpha} e^{-sT} R; \; s \sim \frac{d}{dt} \tag{10}$$

where V is the impoundment water volume (m^3 day^{-1}), R is the rainfall (mm day^{-1}), T is the pure time delay between R and an initial V response, where present (given in number of day intervals), α is the parameter capturing the rate of impoundment depletion or residence time of response (day^{-1}), β is the parameter capturing the magnitude of volume gain (m^3 V, mm^{-1} R), t is time in 1 day periods, and s is the Laplace operator. Full details of this method are given in Young (2015).

Physical interpretation of the behaviour is made after first calculating the dynamic response characteristics (DRCs) of the observed dynamics derived from transfer function parameters α and β. These DRCs include the time constant (TC) or residence time of response (here residence time of the storage following a unit input of rainfall):

$$TC = \frac{\Delta t}{\alpha} \tag{11}$$

where Δt is the time-step (days).

RESULTS AND DISCUSSION

Analysis of infiltration-excess overland-flow inputs to each impoundment

The accumulated infiltration-excess overland flow input to each impoundment (from the surrounding 'rainwater harvesting' catchment) over the 421 days of measurements was 10,690 and 6,630 m^3 for the cattle and control impoundments, respectively (Table 1). This is 1.7 and 1.6 times the maximum water storage capacity of the respective impoundments.

The inputs to the two impoundments were characterized by very short, discrete events of infiltration-excess overland flow, contrasting with the continuous losses to evaporation, leakage and cattle drinking (Figure 2(a)). The 421 days of the study period received 616.5 mm of rainfall. This was distributed in daily events ranging in size from 0.1 to 60 mm day^{-1}. The largest 10% of rainfall events accounted for 42.9% of total rainfall amount, reflecting the typical asymmetric distribution of rainfall in drylands (Knapp *et al.* 2015; Magliano *et al.* 2015b).

Table 1 | Total water inflows by infiltration-excess overland flow ('water harvested'), total water losses and their components (including percent of total loss) of open water evaporation from the impoundment, leakage through the floor of the impoundment and direct consumption of impoundment water by cattle

	Cattle		Control	
	m^3	%	m^3	%
Water harvested	10,690		6,630	
Water losses	10,673		6,835	
Storage change	17		−205	
Evaporation	4,666	43	2,392	35
Basal pond leakage	1,549	15	4,443	65
Consumption	4,458	42		

The smaller rainfall events (<20 mm day^{-1}) seemed to generate much less infiltration-excess overland flow and so an increase in storage compared to the larger events (>30 mm day^{-1}). Indeed both impoundments reached their maximum storage capacity only once in the study period, after a 60 mm rainfall event that was swiftly followed by a 45 mm event. Both impoundments were completely filled once on 13th February 2011. On this day, the maximum water depth was 2.19 m in the cattle impoundment, while water volume and water surface area were 6,140 m^3 and 6,134 m^2, respectively. In the control impoundment, the maximum water depth was 2.08 m, maximum water volume was 4,102 m^3 and maximum water surface area was 4,543 m^2.

To demonstrate the role of the event characteristics on preferential impoundment filling, the overland inflows into the impoundments were explored at a higher temporal resolution (30 min) during two example rainfall events of similar total event size but different short term intensity characteristics (Figure 3). In the cattle impoundment, the high-intensity event generated five times more overland flow than the low-intensity longer duration event (3,742 vs. 710 m^3); and 10 times more in the control impoundment (2,937 vs. 281 m^3; Figure 3).

Examining the full record of daily events, the cattle impoundment had 16 discrete input events while the control

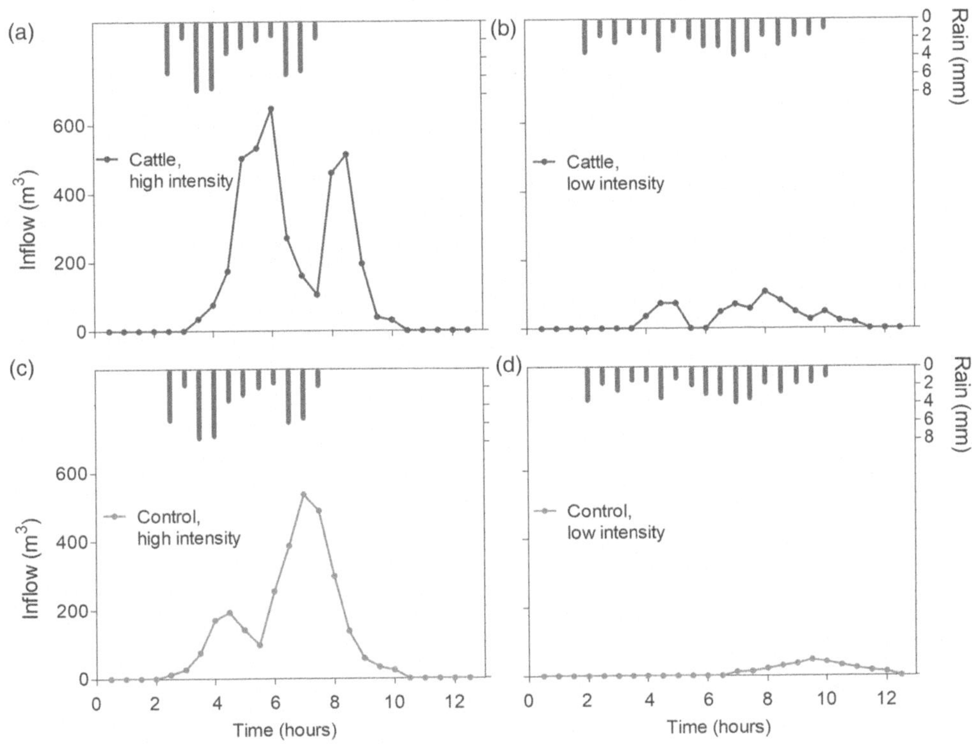

Figure 3 | Inflows recorded every 30 min in the cattle- (a) and (b) and control- (c) and (d) impoundments during example high-intensity (a) and (c) and low-intensity (b) and (d) rainfall events. High-intensity rainfall corresponds to a 45 mm event size with a weighted average intensity of 5.3 mm h^{-1}. Low-intensity rainfall corresponds to a 40 mm event size with a weighted average intensity of 2.8 mm h^{-1}.

had 13. Critically, the minimum daily rainfall event able to generate an infiltration-excess overland flow input to an impoundment was only 4 mm day^{-1} in the cattle impoundment and a larger 20 mm day^{-1} in the control. These observations may indicate that the surface of the catchment draining to the cattle impoundment was naturally less permeable than that of the control impoundment or that the cattle in the former catchment have more extensively compacted the ground enhancing the proportion of infiltration-excess overland flow relative to catchment-wide infiltration. This is consistent with the observations of long compacted cattle trails directed towards drinking impoundments seen in other regions (Ganskopp *et al.* 2000; Ganskopp & Bohnert 2009; Chillo *et al.* 2015).

Analysis of consumption, evaporation and leakage losses from each impoundment

The analysis of the storage dynamics showed that total water losses from the impoundments to all processes over the 421-day monitoring period were 10,673 and 6,835 m^3 from the cattle and control impoundments, respectively (Table 1). The difference between impoundment storage at the start and end of the 421-day period, for both impoundments, was negligible. Consequently, application of Equations (6)–(8) indicates that consumption by cattle represents 42% of the water input to the impoundment where cattle walk inside the impoundment to drink, with a similar amount lost to evaporation (43%), plus 15% lost through infiltration ('leakage') into the impoundment floor. In marked contrast, infiltration through the impoundment floor of the impoundment with no cattle access (or consumption) amounted to 65% of the water entering the impoundment, with a further 35% lost to evaporation.

Depletion of impoundment storage due to the combined losses (expressed by reductions in impoundment storage within Figure 2(a)), followed a clear exponential relationship ($R^2 = 0.82$ and 0.96 for cattle and control impoundment, respectively; Figure 4). The control

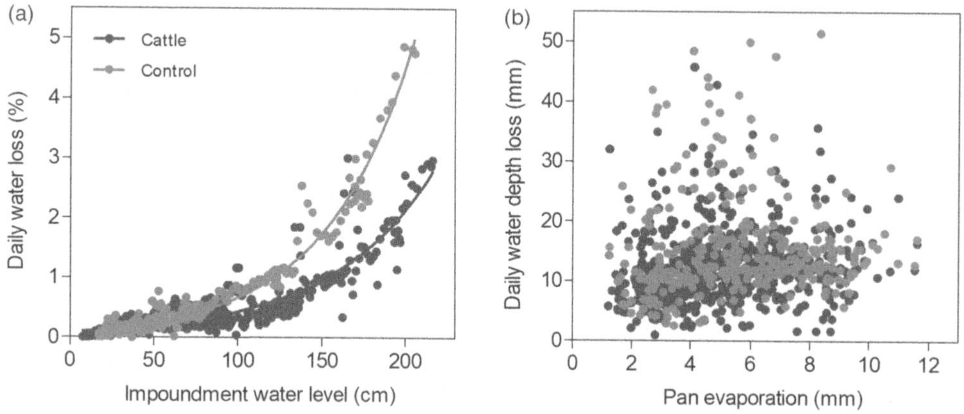

Figure 4 | Relationship between daily water loss (by all processes affecting that impoundment), expressed as a percentage of impoundments maximum storage capacity versus water level of both impoundments.

impoundment lost its water at a higher rate. For example, when the impoundments were full of water, daily losses were 4–5 and 2–3% of stored volume, for the control and cattle impoundments, respectively. Expressed differently, the cattle impoundment showed smaller variations in storage that were maintained closer to maximum storage on average (i.e. 20.4 ± 50.9% coefficient of variation, compared to 9.8 ± 102.0% for the control impoundment; Figure 2(a) inset).

At both impoundments the estimate of the evaporation was at its highest when the storage was at its greatest (Figure 2(a) and 2(b)), as the lateral extent of the impoundment surface was at its greatest at these times. Similarly infiltration (or 'leakage') into the impoundment floor was greatest at times of deepest water, as these were times of greatest hydraulic head driving the water into the impoundment floor. The rate of water loss into the soil below the control impoundment (relative to observed max storage) was, however, twice that beneath the impoundment where cattle were allowed to access. Indeed, 62% of the storage in the control impoundment was lost by infiltration in just one month (i.e. 13th February–15th March 2011). Either the impoundment floor of the 'cattle impoundment' has naturally less permeable soils, or trampling of the impoundment floor by cattle when they access to drink has reduced the permeability of the impoundment floor (so reducing leakage). Indeed wet soils beneath impoundments are likely to be particularly sensitive to cattle trampling (Belsky *et al.* 1999; Lebron *et al.* 2007; Fernández *et al.* 2015; Magliano *et al.* 2017b).

Systems (transfer function) modelling of input-storage dynamics

Application of the RIVC identification algorithm to the rainfall to storage dynamics of the two impoundments was shown to be able to capture more than 90% of the variance in the daily impoundment volume, m^3 (Figure 5; Table 2), thereby permitting comparison of the derived response characteristics between the two impoundments. Purely linear, first-order transfer function model parameters (see Equation (10)) were first identified for the cattle impoundment and a Nash–Sutcliffe simulation efficiency (R_t^2) of 0.875 was observed. The generation of infiltration-excess overland flow is sensitive to the wetness of the surface of the catchment. This may be represented by a single 'memory term' based on the rainfall history. Within this study, the memory term of the 'Bedford-Ouse Sub-Model', BOSM (see Equations (5) and (6) in (Chappell *et al.* 2006) was applied to the rainfall data, with R_{eff} replacing R in Equation (10). By including this one additional term, simulation efficiency increased to 0.945 (Table 2). It is worth mentioning that adding additional parameters can introduce uncertainty to simulation where the information content of the observations is insufficient to warrant the addition of further parameters. This was evaluated with the calculation of the Young Information Criterion, YIC (Young 2015). In the case of the cattle-affected impoundment, the addition of the single BOSM term shifted the value of the YIC performance measure from −8.584 to −10.391 (Table 2). A more negative number indicates

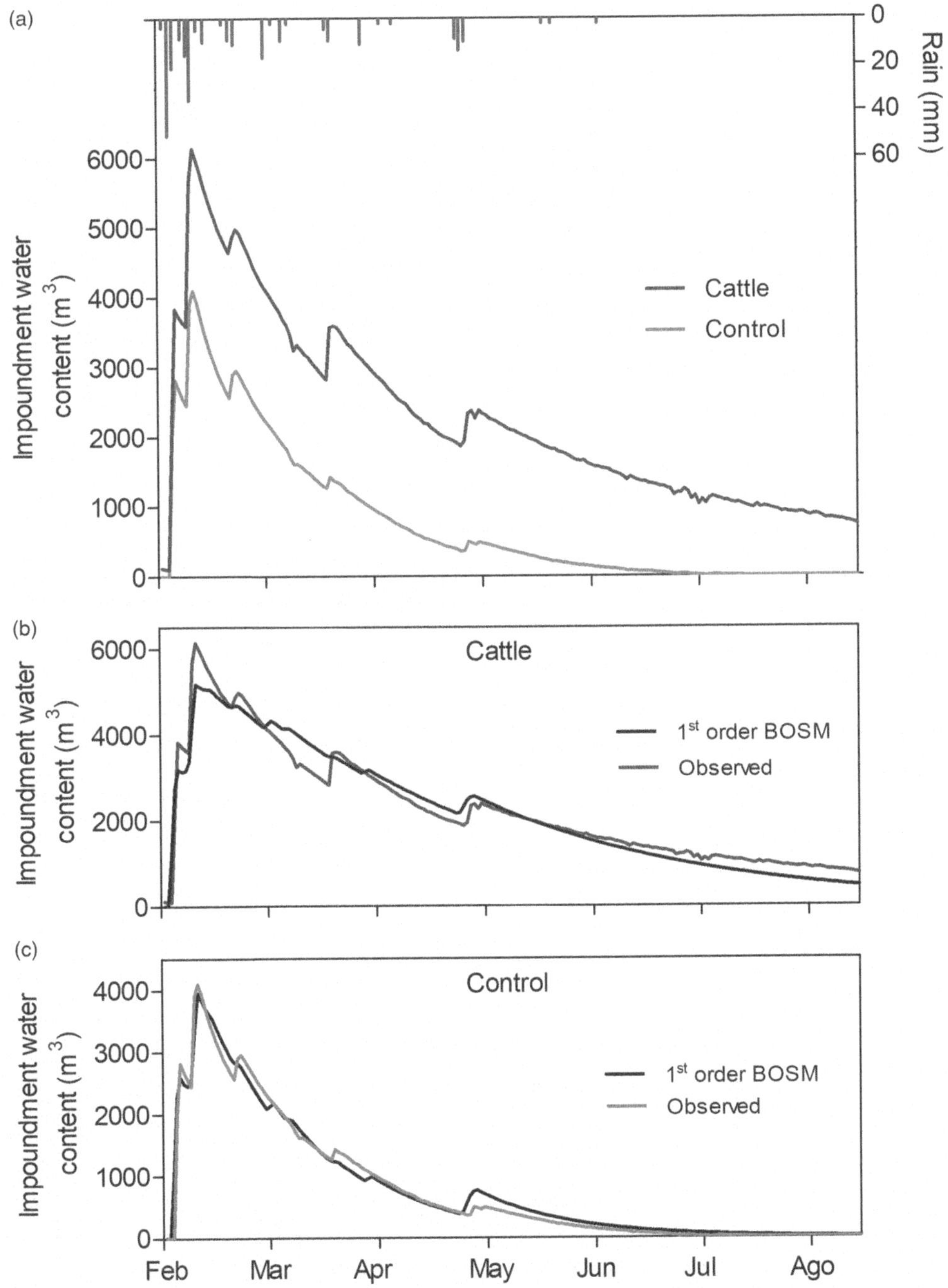

Figure 5 | Observed rainfall and impoundment volume: (a) together with the RIVC-simulated cattle- (b) and control- (c) impoundment volume.

that the use of a more slightly complex structure (i.e. adding one additional term) is warranted by the dynamics in the observed data.

The derived residence time of the response of the storage for a unit input of rainfall to the catchment (Equation (11)) was 62 days for the impoundment affected by cattle access. In contrast, the residence time for the storage in control impoundment was only 25 days (Figure 5(c) and Table 2). Thus, despite the consumption of water by cattle drinking, the less permeable bed of the cattle-affected impoundment retained the water in the impoundment for longer. This greater retention may have been an unintended

Table 2 | The identified transfer function parameters of BOSM nonlinearity term and pure time delay with the derived dynamic response characteristic of the time constant or residence time of response for each impoundment, together with measures of simulation performance (R_t^2, YIC)

	Cattle	Control
BOSM rainfall memory (days)	2	2
Pure time delay, τ (days)	0	0
Measures of simulation performance		
R_t^2	0.94469	0.98272
YIC	−10.3911	−11.888
Time constant or residence time of response (days)	62.1183	25.0948

The identification routine of the RIVC (Refined Instrumental Variable in Continuous-time) algorithm of Young (2015) was applied to the 421 days of daily rainfall and impoundment volume observations.

beneficial effect of the cattle trampling the floor of the impoundment when accessing to drink.

CONCLUSIONS

Rainwater harvesting remains essential for cattle production in the Dry Chaco rangelands of Argentina. This pioneering study into the hydrological functioning of a typical harvesting-impoundment system used for watering cattle against a similar physical system, except for cattle exclusion from the impoundment, produced the following key conclusions:

- The impoundment functioning (cattle-affected and control) is characterised by very short filling events (from infrequent rainfall events) with continuous losses of storage to evaporation and leakage (and also consumption in the cattle-affected impoundment);
- Most inflow is attributed to rainfall events of greater short-term intensity, due to the greater generation of infiltration-excess overland flow rather than infiltration;
- Some 43% of the storage was lost to open water evaporation in the cattle affected impoundment, and a similar 35% from the control impoundment;
- Using the value of 0.05 m^3 per cow multiplied by 300 cows for water consumption from within the cattle-affected impoundment, this left only 15% of the water balance to loss by leakage through the impoundment floor. In considerable contrast, leakage through the impoundment floor amounted to 65% of the stored water in the impoundment without cattle. This greater leakage may be attributed tentatively to the absence of the beneficial effect of cattle trampling reducing the saturated hydraulic conductivity of the impoundment floor;
- Calculation of the water balance components involves uncertainty, particularly the estimates of cattle water consumption and open water evaporation, with the resultant impact on the estimate of impoundment leakage. Future studies on the hydrological functioning of these rainwater harvesting and storage systems should focus on quantifying and reducing the uncertainties in these three key variables;
- The systems modelling allowed quantification of the residence time of storage to unit inputs of rainfall, showing that the impoundment used for cattle drinking counterintuitively maintained storage longer (62 days residence time) than that of the impoundment without cattle ingress (25 days residence time). The beneficial effect of cattle trampling on the saturated impoundment floor to reducing leakage being a plausible first interpretation.

If further studies (with greater constraints on uncertainty in comparison to this pioneering study) similarly show the beneficial effect of cattle trampling on reducing leakage through the impoundment floor, this would have management implications for cattle ranchers in the Dry Chaco rangelands of Argentina. Cattle acting unconsciously as beneficial 'ecosystem engineers' in reducing impoundment leakage, in addition to their acknowledged impact of enhancing impoundment inflows by trampling effects in the contributory catchment, is a finding that cattle ranchers should be aware of. Semi-arid regions across the world are increasingly affected by livestock grazing (Schlesinger *et al.* 1990; Reynolds *et al.* 2007). It is critical that hydrologists undertake experimental studies to quantify both beneficial and negative impacts of these grazers on the sustainability of the natural environment and the food producing systems.

ACKNOWLEDGEMENTS

We thank Ann Kretzschmar, Keith Beven and Raul Giménez for their valuable suggestions. We also thank

George Castellanos and Ricardo Páez for their help with field measurements and analysis. We also thank Martin Ibarra, Alejandro Sanchez and Nicolas Rios Centeno (Ser Beef S.A.) for providing field information and for allowing us to work in the field. This work was funded by the ANPCyT-Argentina (PICT 2016-4136, PICT 2014-2790 and PICT 2013-2973), the CONICET-Argentina (PIP 112-201501-00609) and the Universities UK International (UUKi) Rutherford Fund Strategic Partner Grants (award reference RF-2018-78). Support from NERC NE/R004722/1 covered Nick Chappell's input to the project. San Luis University (Argentina), Lancaster University (United Kingdom) and San Luis Applied Maths Institute (CONICET; Argentina) provided laboratory spaces and workplaces.

REFERENCES

Adham, A., Riksen, M., Ouessar, M. & Ritsema, C. J. 2016 A methodology to assess and evaluate rainwater harvesting techniques in (semi-) arid regions. *Water* **8**, 1–23.

Aguilera, M. 2003 With clear goals. The agricultural station of San Luis. In: *Livestock Use of the Natural Grasslands of San Luis* (M. Aguilera & J. Panigatti, eds). pp. 40.

Aguilera, M. O., Steinaker, D. F. & Demaria, M. R. 2003 Runoff and soil loss in undisturbed and roller-seeded shrublands of semiarid Argentina. *J. Range Manag.* **56**, 227–233.

Allen, R. G., Pereira, L. S., Raes, D. & Smith, M. 1998 Crop evapotranspiration-Guidelines for computing crop water requirements. FAO Irrigation and Drainage Paper 56. FAO, Rome.

Bailey, D. W. 2005 Identification and creation of optimum habitat conditions for livestock. *Rangel. Ecol. Manag.* **58**, 109–118.

Belsky, A. J., Matzke, A. & Uselman, S. 1999 Survey of livestock influences on stream and riparian ecosystems in the western United States. *J. Soil Water Conserv.* **54**, 419–431.

Boers, T. M. & Ben-Asher, J. 1982 A review of rainwater harvesting. *Agric. Water Manag.* **5**, 145–158.

Cabrera, A. L. 1976 *Argentine Encyclopedia of Agriculture and Gardening: Argentine Phytogeographic Regions*. Acme, Buenos Aires, Argentina.

Chappell, N. A., Tych, W., Chotai, A., Bidin, K., Sinun, W. & Chiew, T. H. 2006 BARUMODEL: combined data based mechanistic models of runoff response in a managed rainforest catchment. *For. Ecol. Manage.* **224**, 58–80.

Chillo, V., Ojeda, R. A., Anand, M. & Reynolds, J. F. 2015 A novel approach to assess livestock management effects on biodiversity of drylands. *Ecol. Indic.* **50**, 69–78.

Critchley, W., Siegert, K., Chapman, C. & Finkel, M. 1991 *Water Harvesting. A Manual for the Design and Construction of Water Harvesting Schemes for Plant Production*. Scientific Publishers, Rome, Italy.

Denison, J. A. & Wotshela, L. 2012 An overview of indigenous, indigenised and contemporary water harvesting and conservation practices in South Africa. *Irrig. Drain.* **61**, 7–23.

Evenari, M., Shanan, L. & Tadmor, N. 1971 *The Negev. The Challenge of a Desert*, 2nd edn. Harvard University Press, USA.

Fernández, P. L., Alvarez, C. R. & Taboada, M. A. 2015 Topsoil compaction and recovery in integrated no-tilled crop–livestock systems of Argentina. *Soil Till. Res.* **153**, 86–94.

Fox, P. & Rockström, J. 2000 Water-harvesting for supplementary irrigation of cereal crops to overcome intra-seasonal dry-spells in the Sahel. *Phys. Chem. Earth B Hydrol. Oceans Atmos.* **25**, 289–296.

Ganskopp, D. 2001 Manipulating cattle distribution with salt and water in large arid-land pastures: a GPS/GIS assessment. *Appl. Anim. Behav. Sci.* **73**, 251–262.

Ganskopp, D. C. & Bohnert, D. W. 2009 Landscape nutritional patterns and cattle distribution in rangeland pastures. *Appl. Anim. Behav. Sci.* **116**, 110–119.

Ganskopp, D., Cruz, R. & Johnson, D. 2000 Least-effort pathways?: a GIS analysis of livestock trails in rugged terrain. *Appl. Anim. Behav. Sci.* **68**, 179–190.

García, A. G., Di Bella, C. M., Houspanossian, J., Magliano, P. N., Jobbágy, E. G., Posse, G., Fernández, R. J. & Nosetto, M. D. 2017 Patterns and controls of carbon dioxide and water vapor fluxes in a dry forest of central Argentina. *Agric. For. Meteorol.* **247**, 520–532.

George, M. R., Larsen, R. E., McDougald, N. K., Tate, K. W., Gerlach, J. D. & Fulgham, K. O. 2004 Cattle grazing has varying impacts on stream-channel erosion in oak woodlands. *Calif. Agric.* **58**, 138–143.

Glendenning, C. J., Van Ogtrop, F. F., Mishra, A. K. & Vervoort, R. W. 2012 Balancing watershed and local scale impacts of rain water harvesting in India – a review. *Agric. Water Manag.* **107**, 1–13.

Huxman, T. E., Snyder, K. A., Tissue, D., Leffler, A. J., Ogle, K., Pockman, W. T., Sandquist, D. R., Potts, D. L. & Schwinning, S. 2004 Precipitation pulses and carbon fluxes in semiarid and arid ecosystems. *Oecologia* **141**, 254–268.

Iriondo, M. 1993 Geomorphology and late quaternary of the Chaco (South America). *Geomorphology* **7**, 289–303.

Jobbágy, E. G., Nosetto, M. D., Santoni, C. S. & Baldi, G. 2008 The ecohydrological challenge of woody-herbaceous transitions in the Chaco-Pampas plains. *Ecol. Austral.* **18**, 305–322.

Karlin, M. S., Karlin, U. O., Coirini, R. O., Reati, G. J. & Zapata, R. M. 2013 *The Arid Chaco*. Universidad Nacional de Córdoba, Córdoba, Argentina, 420 pp.

Knapp, A. K., Hoover, D. L., Wilcox, K. R., Avolio, M. L., Koerner, S. E., La Pierre, K. J., Loik, M. E., Luo, Y., Sala, O. E. & Smith, M. D. 2015 Characterizing differences in precipitation regimes of extreme wet and dry years: implications for climate change experiments. *Glob. Change Biol.* **21** (7), 2624–2633.

Kunst, C. R., Ledesma, M., Basan Nickish, G., Angella, D. & Godoy, J. P. 2003 Roller chopping and water infiltration into the soil in the western Chaco (Argentina). *Rev. Invest. Agr.* **32**, 105–126.

Lavee, H., Poesen, J. & Yair, A. 1997 Evidence of high efficiency water-harvesting by ancient farmers in the Negev Desert, Israel. *J. Arid Environ.* **35**, 341–348.

Lebron, I., Madsen, M., Chandler, D., Robinson, D., Wendroth, O. & Belnap, J. 2007 Ecohydrological controls on soil moisture and hydraulic conductivity within a pinyon-juniper woodland. *Water Resour. Res.* **43**, 1–15.

Macchi, L. & Grau, H. R. 2012 Piospheres in the dry Chaco. Contrasting effects of livestock puestos on forest vegetation and bird communities. *J. Arid Environ.* **87**, 176–187.

Magliano, P. N., Breshears, D. D., Fernández, R. J. & Jobbágy, E. G. 2015a Rainfall intensity switches ecohydrological runoff/runon redistribution patterns in dryland vegetation patches. *Ecol. Appl.* **25**, 2094–2100.

Magliano, P. N., Fernández, R. J., Mercau, J. L. & Jobbágy, E. G. 2015b Precipitation event distribution in central Argentina: spatial and temporal patterns. *Ecohydrology* **8**, 94–104.

Magliano, P. N., Murray, F., Baldi, G., Aurand, S., Páez, R. A., Harder, W. & Jobbágy, E. G. 2015c Rainwater harvesting in Dry Chaco: regional distribution and local water balance. *J. Arid Environ.* **123**, 93–102.

Magliano, P. N., Fernández, R. J., Giménez, R., Marchesini, V. A., Páez, R. A. & Jobbágy, E. G. 2016 Changes in water fluxes partition in the Arid Chaco caused by the replacement of forest by pastures. *Ecolog. Austral.* **26**, 95–106.

Magliano, P., Giménez, R., Houspanossian, J., Páez, R., Nosetto, M., Fernández, R. & Jobbágy, E. 2017a Litter is more effective than forest canopy reducing soil evaporation in Dry Chaco rangelands. *Ecohydrology* **10**, e(1879).

Magliano, P. N., Fernández, R. J., Florio, E. L., Murray, F. & Jobbágy, E. G. 2017b Soil physical changes after conversion of woodlands to pastures in Dry Chaco rangelands (Argentina). *Rangel. Ecol. Manage.* **70**, 225–229.

Marchesini, V. A., Fernández, R. J., Reynolds, J. F., Sobrino, J. A. & Di Bella, C. M. 2015 Changes in evapotranspiration and phenology as consequences of shrub removal in dry forests of central Argentina. *Ecohydrology* **8**, 1304–1311.

New, M., Lister, D., Hulme, M. & Makin, I. 2002 A high-resolution data set of surface climate over global land areas. *Clim. Res.* **21**, 1–25.

Newman, B. D., Wilcox, B. P., Archer, S. R., Breshears, D. D., Dahm, C. N., Duffy, C. J., McDowell, N. G., Phillips, F. M., Scanlon, B. R. & Vivoni, E. R. 2006 Ecohydrology of water-limited environments: a scientific vision. *Water Resour. Res.* **42**, 1–15.

Oweis, T. & Hachum, A. 2009 Water harvesting for improved rainfed agriculture in the dry environments. In: *Rainfed Agriculture: Unlocking the Potential* (S. P. Wani, ed.). CAB International, London, UK, pp. 164–182.

Oyarzabal, M., Clavijo, J., Oakley, J., Biganzoli, F., Tognetti, P., Barberis, I., Maturo, H., Aragon, R., Campanello, P., Prado, D., Oesterheld, M. & Leon, R. J. C. 2018 Vegetation units of Argentina. *Ecolog. Austral.* **28**, 40–63.

Pandey, D. N., Gupta, A. K. & Anderson, D. M. 2003 Rainwater harvesting as an adaptation to climate change. *Curr. Sci.* **85**, 46–59.

Peña Zubiate, C. A., Anderson, D. L., Demmi, M. A., Saenz, J. L. & D'Hiriart, A. 1998 *Soils and Vegetation of San Luis Province.* Instituto Nacional de Tecnología Agropecuaria, San Luis, Argentina.

Pennington, T. R., Prado, D. E. & Pendry, C. A. 2000 Neotropical seasonally dry forests and quaternary vegetation changes. *J. Biogeogr.* **27**, 261–273.

Reynolds, J. F., Stafford Smith, D. M., Lambin, E. F., Turner Ii, B. L., Mortimore, M., Batterbury, S. P. J., Downing, T. E., Dowlatabadi, H., Fernández, R. J., Herrick, J. E., Huber-Sannwald, E., Jiang, H., Leemans, R., Lynam, T., Maestre, F. T., Ayarza, M. & Walker, B. 2007 Ecology: global desertification: building a science for dryland development. *Science* **316**, 847–851.

Richardson, D. M. & Rejmánek, M. 2011 Trees and shrubs as invasive alien species – a global review. *Divers. Distrib.* **17** (5), 788–809.

Robinson, T. P., Wint, G. W., Conchedda, G., Van Boeckel, T. P., Ercoli, V., Palamara, E., Cinardi, G., D'Aietti, D., Hay, S. I. & Gilbert, M. 2014 Mapping the global distribution of livestock. *PloS one* **9**, e96084.

Schimel, D. S. 2010 Drylands in the earth system. *Science* **327**, 418–419.

Schlesinger, W. H., Reynolds, J. F., Cunningham, G. L., Huenneke, L. F., Jarrell, W. M., Virginia, R. A. & Whitford, W. G. 1990 Biological feedbacks in global desertification. *Science* **247**, 1043–1048.

Schlink, A., Nguyen, M. & Viljoen, G. 2010 Water requirements for livestock production: a global perspective. *Rev. Sci. Tech.* **29**, 603–619.

Steinaker, D. F., Jobbágy, E. G., Martini, J. P., Arroyo, D. N., Pacheco, J. L. & Marchesini, V. A. 2016 Vegetation composition and structure changes following roller-chopping deforestation in central Argentina woodlands. *J. Arid Environ.* **133**, 19–24.

Taylor, C. J., Pedregal, D. J., Young, P. C. & Tych, W. 2007 Environmental time series analysis and forecasting with the Captain toolbox. *Environ. Model. Softw.* **22**, 797–814.

Tripaldi, A., Zárate, M. A., Forman, S. L., Badger, T., Doyle, M. E. & Ciccioli, P. 2013 Geological evidence for a drought episode in the western Pampas (Argentina, South America) during the early–mid 20th century. *Holocene* **23**, 1731–1746.

UNEP 2009 *Rainwater Harvesting: A Lifeline for Human Well-Being.* United Nations Environment Programme, Stockholm, Sweden.

Young, P. C. 2015 Refined instrumental variable estimation: maximum likelihood optimization of a unified Box–Jenkins model. *Automatica* **52**, 35–46.

First received 28 November 2018; accepted in revised form 16 September 2019. Available online 21 October 2019